D1206928

Introduction to Graph Theory

Fourth edition

Introduction to Graph Theory

Fourth edition

Robin J. Wilson

An imprint of **Pearson Education**

Harlow, England · London · New York · Reading, Massachusetts · San Francisco · Toronto · Don Mills, Ontario · Sydney
Tokyo · Singapore · Hong Kong · Seoul · Taipei · Cape Town · Madrid · Mexico City · Amsterdam · Munich · Paris · Milan

Pearson Education Limited
Edinburgh Gate
Harlow
Essex CM20 2JE
England

and Associated Companies throughout the world

Visit us on the World Wide Web at:
www.pearsoned.co.uk

First published by Oliver & Boyd, 1972
Second edition published by Longman Group Ltd, 1979
Third edition, 1985
Fourth edition, 1996

British Library Cataloguing in Publication Data
A catalogue record for this book is available from the British Library

ISBN 978-0-582-24993-6

Library of Congress Cataloging-in-Publication Data
A catalog record for this book is available from the Library of Congress

14 13 12
09 08 07

Set by 8 in 10 on 12pt Times
Printed in Malaysia, PP

Contents

Go forth, my little book! pursue thy way!
Go forth, and please the gentle and the good.
William Wordsworth

Preface to the fourth edition

In recent years, graph theory has established itself as an important mathematical tool in a wide variety of subjects, ranging from operational research and chemistry to genetics and linguistics, and from electrical engineering and geography to sociology and architecture. At the same time it has also emerged as a worthwhile mathematical discipline in its own right.

In view of this, there is a need for an inexpensive introductory text on the subject, suitable both for mathematicians taking courses in graph theory and also for non-specialists wishing to learn the subject as quickly as possible. It is my hope that this book goes some way towards filling this need. The only prerequisites to reading it are a basic knowledge of elementary set theory and matrix theory, although a further knowledge of abstract algebra is needed for more difficult exercises.

The contents of this book may be conveniently divided into four parts. The first of these (Chapters 1–4) provides a basic foundation course, containing definitions and examples of graphs, connectedness, Eulerian and Hamiltonian paths and cycles, and trees. This is followed by two chapters (Chapters 5 and 6) on planarity and colouring, with special reference to the four-colour theorem. The third part (Chapters 7 and 8) deals with the theory of directed graphs and with transversal theory, with applications to critical path analysis, Markov chains and network flows. The book ends with a chapter on matroids (Chapter 9), which ties together material from the previous chapters and introduces some recent developments.

Throughout the book I have attempted to restrict the text to basic material, using exercises as a means for introducing less important material. Of the 250 exercises, some are routine examples designed to test understanding of the text, while others will introduce you to new results and ideas. You should read each exercise, whether or not you work through it in detail. Difficult exercises are indicated by an asterisk.

I have used the symbol // to indicate the end of a proof, and bold-face type is used for definitions. The number of elements in a set S is denoted by $|S|$, and the empty set is denoted by $\varnothing$.

A substantial number of changes have been made in this edition. The text has been revised throughout, and some terminology has been changed to fit in with current usage. In addition, solutions are given for some of the exercises; these exercises are indicated by the symbol [s] next to the exercise number. Several changes have arisen as a result of comments by a number of people, and I should like to take this opportunity of thanking them for their helpful remarks.

Finally, I wish to express my thanks to my former students, but for whom this book would have been completed a year earlier, to Mr William Shakespeare and others for their apt and witty comments at the beginning of each chapter, and most of all to my wife Joy for many things that have nothing to do with graph theory.

R.J.W.
May 1995
The Open University

Chapter 1

Introduction

The last thing one discovers in
writing a book is what to put first.
Blaise Pascal

In this introductory chapter we provide an intuitive background to the material that we present more formally in later chapters. Terms that appear here in bold-face type are to be thought of as descriptions rather than as definitions. Having met them here in an informal setting, you should find them more familiar when you meet them later. So read this chapter quickly, and then forget all about it!

1 What is a graph?

We begin by considering Figs. 1.1 and 1.2, which depict part of a road map and part of an electrical network.

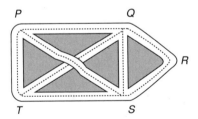

Fig. 1.1

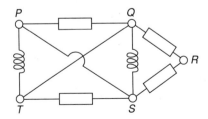

Fig. 1.2

Either of these situations can be represented diagrammatically by means of points and lines, as in Fig. 1.3. The points P, Q, R, S and T are called **vertices**, the lines are called **edges**, and the whole diagram is called a **graph**. Note that the intersection of the lines PS and QT is not a vertex, since it does not correspond to a cross-roads or to the meeting of two wires. The **degree** of a vertex is the number of edges with that vertex as an end-point; it corresponds in Fig. 1.1 to the number of roads at an intersection. For example, the degree of the vertex Q is 4.

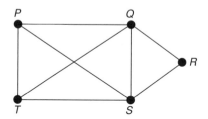

Fig. 1.3

The graph in Fig. 1.3 can also represent other situations. For example, if P, Q, R, S and T represent football teams, then the existence of an edge might correspond to the playing of a game between the teams at its end-points. Thus, in Fig. 1.3, team P has played against teams Q, S and T, but not against team R. In this representation, the degree of a vertex is the number of games played by the corresponding team.

Another way of depicting these situations is by the graph in Fig. 1.4. Here we have removed the 'crossing' of the lines PS and QT by drawing the line PS outside the rectangle $PQST$. The resulting graph still tells us whether there is a direct road from one intersection to another, how the electrical network is wired up, and which football teams have played which. The only information we have lost concerns 'metrical' properties, such as the length of a road and the straightness of a wire.

Thus, a graph is a representation of a set of points and of how they are joined up, and any metrical properties are irrelevant. From this point of view, any graphs that represent the same situation, such as those of Figs. 1.3 and 1.4, are regarded as the same graph.

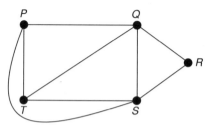

Fig. 1.4

More generally, two graphs are the same if two vertices are joined by an edge in one graph if and only if the corresponding vertices are joined by an edge in the other. Another graph that is the same as the graphs in Figs. 1.3 and 1.4 is shown in Fig. 1.5. Here all idea of space and distance has gone, although we can still tell at a glance which points are joined by a road or a wire.

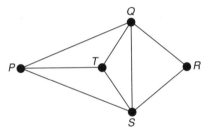

Fig. 1.5

In the above graph there is at most one edge joining each pair of vertices. Suppose now, that in Fig. 1.5 the roads joining Q and S, and S and T, have too much traffic to carry. Then the situation is eased by building extra roads joining these points, and the resulting diagram looks like Fig. 1.6. The edges joining Q and S, or S and T, are called **multiple edges**. If, in addition, we need a car park at P, then we indicate this by drawing an edge from P to itself, called a **loop** (see Fig. 1.7). In this book, a graph may contain loops and multiple edges. Graphs with no loops or multiple edges, such as the graph in Fig. 1.5, are called **simple graphs**.

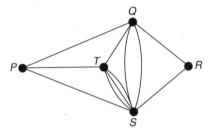

Fig. 1.6

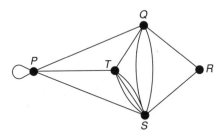

Fig. 1.7

The study of **directed graphs** (or **digraphs,** as we abbreviate them) arises from making the roads into one-way streets. An example of a digraph is given in Fig. 1.8, the directions of the one-way streets being indicated by arrows. (In this example, there would be chaos at T, but that does not stop us from studying such situations!) We discuss digraphs in Chapter 7.

Much of graph theory involves 'walks' of various kinds. A **walk** is a 'way of getting from one vertex to another', and consists of a sequence of edges, one following after another. For example, in Fig 1.5 $P \rightarrow Q \rightarrow R$ is a walk of length 2, and $P \rightarrow S \rightarrow Q \rightarrow T \rightarrow S \rightarrow R$ is a walk of length 5. A walk in which no vertex appears more than once is

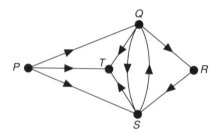

Fig. 1.8

called a **path**; for example, $P \to T \to S \to R$ is a path. A walk of the form $Q \to S \to T \to Q$ is called a **cycle**.

Much of Chapter 3 is devoted to walks with some special property. In particular, we discuss graphs containing walks that include every edge or every vertex exactly once, ending at the initial vertex; such graphs are called **Eulerian** and **Hamiltonian** graphs, respectively. For example, the graph in Figs 1.3–1.5 is Hamiltonian; a suitable walk is $P \to Q \to R \to S \to T \to P$. It is not Eulerian, since any walk that includes each edge exactly once (such as $P \to Q \to R \to S \to T \to P \to S \to Q \to T$) must end at a vertex different from the initial one.

Some graphs are in two or more parts. For example, consider the graph whose vertices are the stations of the London Underground and the New York Subway, and whose edges are the lines joining them. It is impossible to travel from Trafalgar Square to Grand Central Station using only edges of this graph, but if we confine our attention to the London Underground only, then we can travel from any station to any other. A graph that is in one piece, so that any two vertices are connected by a path, is a **connected graph**; a graph in more than one piece is a **disconnected graph** (see Fig. 1.9). We discuss connectedness in Chapter 3.

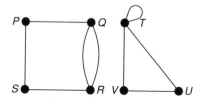

Fig. 1.9

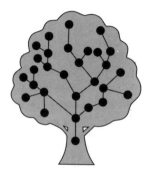

Fig. 1.10

We are sometimes interested in connected graphs with only one path between each pair of vertices. Such graphs are called **trees**, generalizing the idea of a family tree, and are considered in Chapter 4. As we shall see, a tree can be defined as a connected graph containing no cycles (see Fig. 1.10).

Earlier we noted that Fig. 1.3 can be redrawn as in Figs 1.4 and 1.5 so as to avoid crossings of edges. A graph that can be redrawn without crossings in this way is called a **planar graph**. In Chapter 5 we give several criteria for planarity. Some of these involve the properties of 'subgraphs' of the graph in question; others involve the fundamental notion of duality.

Planar graphs also play an important role in colouring problems. In our 'road-map' graph, let us suppose that Shell, Esso, BP, and Gulf wish to erect five garages between them, and that for economic reasons no company wishes to erect two garages at neighbouring corners. Then Shell can build at P, Esso can build at Q, BP can build at S, and Gulf can build at T, leaving either Shell or Gulf to build at R (see Fig. 1.11). However, if Gulf backs out of the agreement, then the other three companies cannot erect the garages in the specified manner.

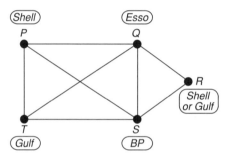

Fig. 1.11

Fig. 1.12

We discuss such problems in Chapter 6, where we try to colour the vertices of a simple graph with a given number of colours so that each edge of the graph joins vertices of different colours. If the graph is planar, then we can always colour its vertices in this way with only four colours – this is the celebrated **four-colour theorem**. Another version of this theorem is that we can always colour the countries of any map with four colours so that no two neighbouring countries share the same colour (see Fig. 1.12).

In Chapter 8 we investigate the celebrated **marriage problem**, which asks under what conditions a collection of girls, each of whom knows several boys, can be married so that each girl marries a boy she knows. This problem can be expressed in the language of 'transversal theory', and is related to problems of finding disjoint paths connecting two given vertices in a graph or digraph.

Chapter 8 concludes with a discussion of network flows and transportation problems. Suppose that we have a transportation network such as in Fig. 1.13, in which P is a factory, R is a market, and the edges of the graph are channels through which goods can be sent. Each channel has a capacity, indicated by a number next to the edge, representing the maximum amount that can pass through that channel. The problem is to determine how much can be sent from the factory to the market.

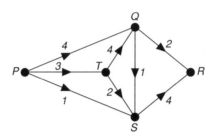

Fig. 1.13

We conclude with a chapter on matroids. This ties together the material of the previous chapters, while satisfying the maxim 'be wise – generalize!' Matroid theory, the study of sets with 'independence structures' defined on them, generalizes both linear independence in vector spaces and some results on graphs and transversals from earlier in the book. However, matroid theory is far from being 'generalization for generalization's sake'. On the contrary, it gives us deeper insight into several graph problems, as well as providing simple proofs of results on transversals that are awkward to prove by more traditional methods. Matroids have played an important role in the development of combinatorial ideas in recent years.

We hope that this introductory chapter has been useful in setting the scene and describing some of the treats that lie ahead. We now embark upon a formal treatment of the subject.

Exercises 1

1.1[s]   Write down the number of vertices, the number of edges, and the degree of each vertex, in:
   (i)   the graph in Fig. 1.3;
   (ii)   the tree in Fig. 1.14.

Fig. 1.14

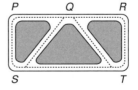

Fig. 1.15

1.2   Draw the graph representing the road system in Fig. 1.15, and write down the number of vertices, the number of edges and the degree of each vertex.

1.3ˢ   Figure 1.16 represents the chemical molecules of methane (CH_4) and propane (C_3H_8).
  (i)   Regarding these diagrams as graphs, what can you say about the vertices representing carbon atoms (C) and hydrogen atoms (H)?
  (ii)   There are two different chemical molecules with formula C_4H_{10}. Draw the graphs corresponding to these molecules.

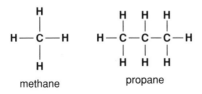

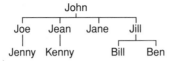

methane   propane

Fig. 1.16

John

Joe   Jean   Jane   Jill

Jenny   Kenny   Bill   Ben

Fig. 1.17

1.4   Draw a graph corresponding to the family tree in Fig. 1.17.

1.5*   Draw a graph with vertices $A, \ldots, M$ that shows the various routes one can take when tracing the Hampton Court maze in Fig. 1.18.

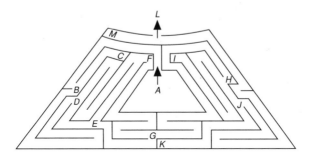

Fig. 1.18

1.6ˢ   John likes Joan, Jean and Jane; Joe likes Jane and Joan; Jean and Joan like each other. Draw a digraph illustrating these relationships between John, Joan, Jean, Jane and Joe.

1.7   Snakes eat frogs and birds eat spiders; birds and spiders both eat insects; frogs eat snails, spiders and insects. Draw a digraph representing this predatory behaviour.

Chapter 2

Definitions and examples

I hate definitions!
Benjamin Disraeli

In this chapter, we lay the foundations for a proper study of graph theory. Section 2 formalizes some of the basic definitions of Chapter 1 and Section 3 provides a variety of examples. In Section 4 we show how graphs can be used to represent and solve three problems from recreational mathematics. More substantial applications are deferred until we have more machinery at our disposal (see Sections 8 and 11).

2 Definitions

A **simple graph** G consists of a non-empty finite set $V(G)$ of elements called **vertices** (or **nodes**), and a finite set $E(G)$ of distinct unordered pairs of distinct elements of $V(G)$ called **edges**. We call $V(G)$ the **vertex set** and $E(G)$ the **edge set** of G. An edge $\{v, w\}$ is said to **join** the vertices v and w, and is usually abbreviated to vw. For example, Fig. 2.1 represents the simple graph G whose vertex set $V(G)$ is $\{u, v, w, z\}$, and whose edge set $E(G)$ consists of the edges uv, uw, vw and wz.

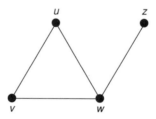

Fig. 2.1

In any simple graph there is at most one edge joining a given pair of vertices. However, many results that hold for simple graphs can be extended to more general objects in which two vertices may have several edges joining them. In addition, we may remove the restriction that an edge joins two *distinct* vertices, and allow **loops** – edges joining a vertex to itself. The resulting object, in which loops and multiple edges are allowed, is called a **general graph** – or, simply, a **graph** (see Fig. 2.2). Thus every simple graph is a graph, but not every graph is a simple graph.

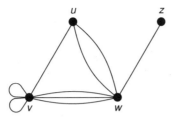

Fig. 2.2

Thus, a **graph** G consists of a non-empty finite set $V(G)$ of elements called **vertices**, and a finite family $E(G)$ of unordered pairs of (not necessarily distinct) elements of $V(G)$ called **edges**; the use of the word 'family' permits the existence of multiple edges[†]. We call $V(G)$ the **vertex set** and $E(G)$ the **edge family** of G. An edge $\{v, w\}$ is said to **join** the vertices v and w, and is again abbreviated to vw. Thus in Fig. 2.2, $V(G)$ is the set $\{u, v, w, z\}$ and $E(G)$ consists of the edges uv, vv (twice), vw (three times), uw (twice), and wz. Note that each loop vv joins the vertex v to itself. Although we sometimes have to restrict our attention to simple graphs, we shall prove our results for general graphs whenever possible.

The language of graph theory is not standard – all authors have their own terminology. Some use the term 'graph' for what we call a simple graph, or for a graph with directed edges, or for a graph with infinitely many vertices or edges; we discuss digraphs in Chapter 7 and infinite graphs in Section 16. Any such definition is perfectly valid, provided that it is used consistently. In this book, *all graphs are finite and undirected, with loops and multiple edges allowed unless specifically excluded.*

Isomorphism

Two graphs G_1 and G_2 are **isomorphic** if there is a one–one correspondence between the vertices of G_1 and those of G_2 such that the number of edges joining any two vertices of G_1 is equal to the number of edges joining the corresponding vertices of G_2. Thus the two graphs shown in Fig. 2.3 are isomorphic under the correspondence $u \leftrightarrow l, v \leftrightarrow m, w \leftrightarrow n, x \leftrightarrow p, y \leftrightarrow q, z \leftrightarrow r$. For many problems, the labels on the vertices are unnecessary and we drop them. We then say that two 'unlabelled graphs' are isomorphic if we can assign labels so that the resulting 'labelled graphs' are

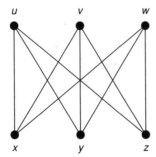
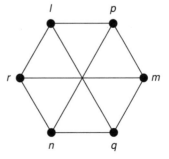

Fig. 2.3

[†] We use the word 'family' to mean a collection of elements, some of which may occur several times; for example, $\{a, b, c\}$ is a set, but (a, a, c, b, a, c) is a family.

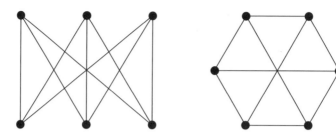

Fig. 2.4

isomorphic. For example, the unlabelled graphs in Fig. 2.4 are isomorphic, since we can label the vertices as in Fig. 2.3.

The difference between labelled and unlabelled graphs becomes more apparent when we try to count them. For example, if we restrict ourselves to graphs with three vertices, then there are, up to isomorphism, eight different labelled graphs but only four unlabelled ones (see Figs 2.5 and 2.6). It is usually clear from the context whether we are referring to labelled or unlabelled graphs.

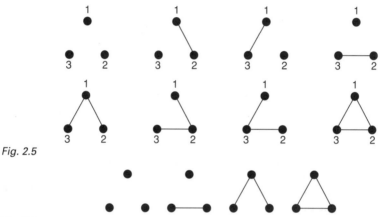

Fig. 2.5

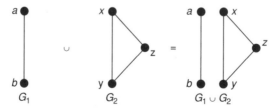

Fig. 2.6

Connectedness

We can combine two graphs to make a larger graph. If the two graphs are $G_1 = (V(G_1), E(G_1))$ and $G_2 = (V(G_2), E(G_2))$, where $V(G_1)$ and $V(G_2)$ are disjoint, then their **union** $G_1 \cup G_2$ is the graph with vertex set $V(G_1) \cup V(G_2)$ and edge family $E(G_1) \cup E(G_2)$ (see Fig. 2.7).

Fig. 2.7

Most all the graphs discussed so far have been 'in one piece'. A graph is **connected** if it cannot be expressed as the union of two graphs, and **disconnected** otherwise. Clearly any disconnected graph G can be expressed as the union of connected graphs, each of which is a **component** of G. For example, a graph with three components is shown in Fig. 2.8.

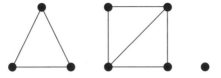

Fig. 2.8

When proving results about graphs in general, we can often obtain the corresponding results for connected graphs and then apply them to each component separately. A table of all the connected unlabelled graphs with up to five vertices is given in Fig. 2.9.

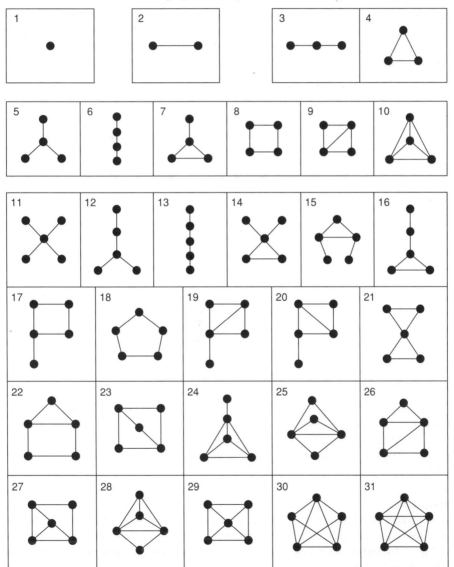

Fig. 2.9

Adjacency

We say that two vertices v and w of a graph G are **adjacent** if there is an edge vw joining them, and the vertices v and w are then **incident** with such an edge. Similarly, two distinct edges e and f are **adjacent** if they have a vertex in common (see Fig. 2.10).

Fig. 2.10

The **degree** of a vertex v of G is the number of edges incident with v, and is written $\deg(v)$; in calculating the degree of v, we usually make the convention that a loop at v contributes 2 (rather than 1) to the degree of v. A vertex of degree 0 is an **isolated vertex** and a vertex of degree 1 is an **end-vertex**. Thus each of the two graphs in Fig. 2.11 has two end-vertices and three vertices of degree 2, while the graph in Fig. 2.12 has one end-vertex, one vertex of degree 3, one of degree 6 and one of degree 8. The **degree sequence** of a graph consists of the degrees written in increasing order, with repeats where necessary. For example, the degree sequences of the graphs in Figs. 2.11 and 2.12 are (1, 1, 2, 2, 2) and (1, 3, 6, 8).

Fig. 2.11

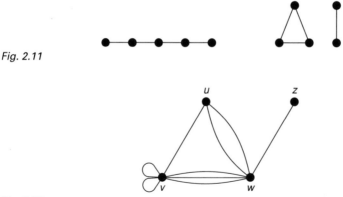

Fig. 2.12

Note that *in any graph the sum of all the vertex-degrees is an even number* – in fact, twice the number of edges, since each edge contributes exactly 2 to the sum. This result, due essentially to Leonhard Euler in 1736, is called the **handshaking lemma**. It implies that if several people shake hands, then the total number of hands shaken must be even – precisely because just two hands are involved in each handshake. An immediate corollary of the handshaking lemma is that *in any graph the number of vertices of odd degree is even.*

Subgraphs

A **subgraph** of a graph G is a graph, each of whose vertices belongs to $V(G)$ and each of whose edges belongs to $E(G)$. Thus the graph in Fig. 2.13 is a subgraph of the graph in Fig. 2.14, but is not a subgraph of the graph in Fig. 2.15, since the latter graph contains no 'triangle'.

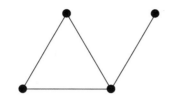

Fig. 2.13

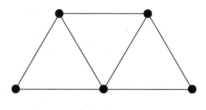

Fig. 2.14

Fig. 2.15

We can obtain subgraphs of a graph by deleting edges and vertices. If e is an edge of a graph G, we denote by $G - e$ the graph obtained from G by deleting the edge e. More generally, if F is any set of edges in G, we denote by $G - F$ the graph obtained by deleting the edges in F. Similarly, if v is a vertex of G, we denote by $G - v$ the graph obtained from G by deleting the vertex v together with the edges incident with v. More generally, if S is any set of vertices in G, we denote by $G - S$ the graph obtained by deleting the vertices in S and all edges incident with any of them. Some examples are shown in Fig. 2.16.

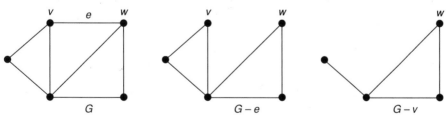

Fig. 2.16

We also denote by $G \backslash e$ the graph obtained by taking an edge e and contracting it – removing it and identifying its ends v and w so that the resulting vertex is incident with those edges (other than e) that were originally incident with v or w. An example is shown in Fig. 2.17.

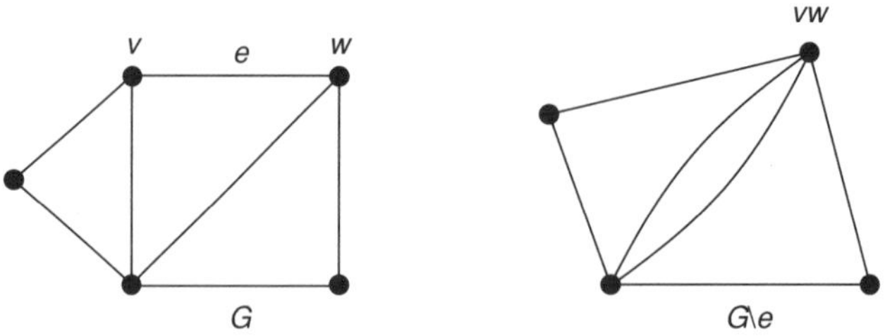

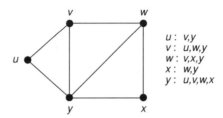

Fig. 2.17

Matrix representations

Although it is convenient to represent a graph by a diagram of points joined by lines, such a representation may be unsuitable if we wish to store a large graph in a computer. One way of storing a simple graph is by listing the vertices adjacent to each vertex of the graph. An example of this is given in Fig. 2.18.

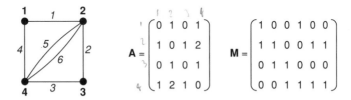

Fig. 2.18

Other useful representations involve matrices. If G is a graph with vertices labelled $\{1, 2, \ldots, n\}$, its **adjacency matrix A** is the $n \times n$ matrix whose ij-th entry is the number of edges joining vertex i and vertex j. If, in addition, the edges are labelled $\{1, 2, \ldots, m\}$, its **incidence matrix M** is the $n \times m$ matrix whose ij-th entry is 1 if vertex i is incident to edge j, and 0 otherwise. Figure 2.19 shows a labelled graph G with its adjacency and incidence matrices.

$$A = \begin{pmatrix} 0 & 1 & 0 & 1 \\ 1 & 0 & 1 & 2 \\ 0 & 1 & 0 & 1 \\ 1 & 2 & 1 & 0 \end{pmatrix} \qquad M = \begin{pmatrix} 1 & 0 & 0 & 1 & 0 & 0 \\ 1 & 1 & 0 & 0 & 1 & 1 \\ 0 & 1 & 1 & 0 & 0 & 0 \\ 0 & 0 & 1 & 1 & 1 & 1 \end{pmatrix}$$

Fig. 2.19

Exercises 2

2.1[s]   Write down the vertex set and edge set of each graph in Fig. 2.3.

2.2   Draw
   (i)   a simple graph,
   (ii)   a non-simple graph with no loops,
   (iii)   a non-simple graph with no multiple edges,
   each with five vertices and eight edges.

2.3ˢ (i) By suitably labelling the vertices, show that the two graphs in Fig. 2.20 are isomorphic.

(ii) Explain why the two graphs in Fig. 2.21 are not isomorphic.

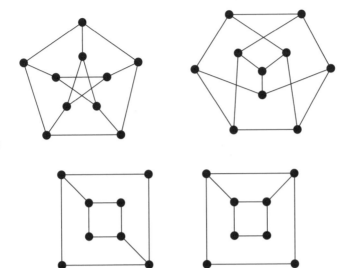

Fig. 2.20

Fig. 2.21

2.4 Classify the following statements as *true* or *false:*

(i) any two isomorphic graphs have the same degree sequence;

(ii) any two graphs with the same degree sequence are isomorphic.

2.5 (i) Show that there are exactly $2^{n(n-1)/2}$ labelled simple graphs on n vertices.

(ii) How many of these have exactly m edges?

2.6ˢ Locate each of the graphs in Fig. 2.22 in the table of Fig. 2.9.

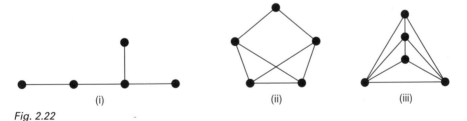

(i) (ii) (iii)

Fig. 2.22

2.7ˢ Write down the degree sequence of each graph with four vertices in Fig. 2.9, and verify that the handshaking lemma holds for each graph.

2.8 (i) Draw a graph on six vertices with degree sequence (3, 3, 5, 5, 5, 5); does there exist a *simple* graph with these degrees?

(ii) How are your answers to part (i) changed if the degree sequence is (2, 3, 3, 4, 5, 5)?

2.9* If G is a simple graph with at least two vertices, prove that G must contain two or more vertices of the same degree.

2.10ˢ Which graphs in Fig. 2.23 are subgraphs of those in Fig. 2.20?

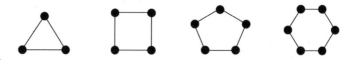

Fig. 2.23

2.11   Let G be a graph with n vertices and m edges, and let v be a vertex of G of degree k and e be an edge of G. How many vertices and edges have $G - e$, $G - v$ and $G\backslash e$?

2.12[s]   Write down the adjacency and incidence matrices of the graph in Fig. 2.24.

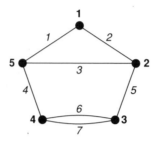

Fig. 2.24

$$\begin{pmatrix} 0 & 1 & 1 & 2 & 0 \\ 1 & 0 & 0 & 0 & 1 \\ 1 & 0 & 0 & 1 & 1 \\ 2 & 0 & 1 & 0 & 0 \\ 0 & 1 & 1 & 0 & 0 \end{pmatrix}$$

Fig. 2.25

$$\begin{pmatrix} 0 & 0 & 1 & 1 & 1 & 1 & 1 & 0 \\ 0 & 1 & 0 & 1 & 0 & 0 & 0 & 1 \\ 0 & 0 & 0 & 0 & 0 & 0 & 0 & 1 \\ 1 & 0 & 1 & 0 & 1 & 0 & 1 & 0 \\ 1 & 1 & 0 & 0 & 0 & 1 & 0 & 0 \end{pmatrix}$$

Fig. 2.26

2.13   (i)   Draw the graph whose adjacency matrix is given in Fig. 2.25.
    (ii)   Draw the graph whose incidence matrix is given in Fig. 2.26.

2.14   If G is a graph without loops, what can you say about the sum of the entries in
    (i)   any row or column of the adjacency matrix of G?
    (ii)   any row of the incidence matrix of G?
    (iii)   any column of the incidence matrix of G?

2.15*   If G is a simple graph with edge-set $E(G)$, the **vector space of** G is the vector space over the field $\mathbf{Z}_2$ of integers modulo 2, whose elements are subsets of $E(G)$. The sum $E + F$ of two subsets E and F is the set of edges in E or F but not both, and scalar multiplication is defined by $1.E = E$ and $0.E = \emptyset$. Show that this defines a vector space over $\mathbf{Z}_2$, and find a basis for it.

3 Examples

In this section we examine some important types of graphs. You should become familiar with them, as they will appear frequently in examples and exercises.

Null graphs

A graph whose edge-set is empty is a **null graph**. We denote the null graph on n vertices by N_n; N_4 is shown in Fig. 3.1. Note that each vertex of a null graph is isolated. Null graphs are not very interesting.

Fig. 3.1

Complete graphs

A simple graph in which each pair of distinct vertices are adjacent is a **complete graph**. We denote the complete graph on n vertices by K_n; K_4 and K_5 are shown in Fig. 3.2. You should check that K_n has $n(n-1)/2$ edges.

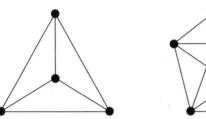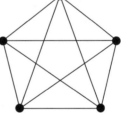

Fig. 3.2

Cycle graphs, path graphs and wheels

A connected graph that is regular of degree 2 is a **cycle graph**. We denote the cycle graph on n vertices by C_n. The graph obtained from C_n by removing an edge is the **path graph** on n vertices, denoted by P_n. The graph obtained from C_{n-1} by joining each vertex to a new vertex v is the **wheel** on n vertices, denoted by W_n. The graphs C_6, P_6 and W_6 are shown in Fig. 3.3.

Fig. 3.3

Regular graphs

A graph in which each vertex has the same degree is a **regular graph**. If each vertex has degree r, the graph is **regular of degree r** or **r-regular**. Of special importance are the **cubic graphs**, which are regular of degree 3; an example of a cubic graph is the **Petersen graph**, shown in Fig. 3.4. Note that the null graph N_n is regular of degree 0, the cycle graph C_n is regular of degree 2, and the complete graph K_n is regular of degree $n-1$.

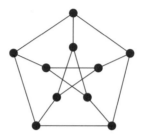

Fig. 3.4

Platonic graphs

Of interest among the regular graphs are the **Platonic graphs**, formed from the vertices and edges of the five regular (Platonic) solids – the tetrahedron, octahedron, cube, icosahedron and dodecahedron (see Fig. 3.5).

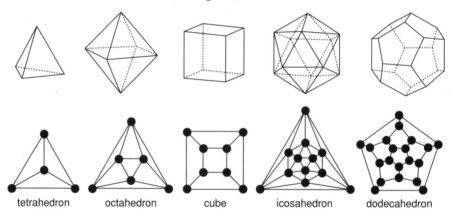

| tetrahedron | octahedron | cube | icosahedron | dodecahedron |

Fig. 3.5

Bipartite graphs

If the vertex set of a graph G can be split into two disjoint sets A and B so that each edge of G joins a vertex of A and a vertex of B, then G is a **bipartite graph** (see Fig. 3.6). Alternatively, a bipartite graph is one whose vertices can be coloured black and white in such a way that each edge joins a black vertex (in A) and a white vertex (in B).

A **complete bipartite graph** is a bipartite graph in which each vertex in A is joined to each vertex in B by just one edge. We denote the bipartite graph with r black vertices and s white vertices by $K_{r,s}$; $K_{1,3}$, $K_{2,3}$, $K_{3,3}$ and $K_{4,3}$ are shown in Fig. 3.7. You should check that $K_{r,s}$ has $r + s$ vertices and rs edges.

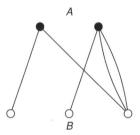

A

B

Fig. 3.6

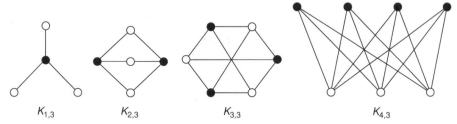

$K_{1,3}$ $K_{2,3}$ $K_{3,3}$ $K_{4,3}$

Fig. 3.7

Cubes

Of special interest among the regular bipartite graphs are the cubes. The **k-cube** Q_k is the graph whose vertices correspond to the sequences $(a_1, a_2, \ldots, a_k)$, where each $a_i = 0$ or 1, and whose edges join those sequences that differ in just one place. Note that the graph of the cube is the graph Q_3 (see Fig. 3.8). You should check that Q_k has 2^k vertices and $k2^{k-1}$ edges, and is regular of degree k.

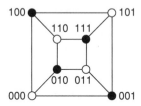

100   101
110 111
010 011
000   001

Fig. 3.8

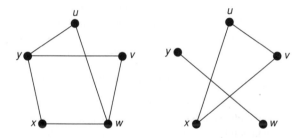

Fig. 3.9

The complement of a simple graph

If G is a simple graph with vertex set $V(G)$, its **complement** $\overline{G}$ is the simple graph with vertex set $V(G)$ in which two vertices are adjacent if and only if they are *not* adjacent in G. For example, Fig. 3.9 shows a graph and its complement. Note that the complement of a complete graph is a null graph, and that the complement of a complete bipartite graph is the union of two complete graphs.

Exercises 3

3.1[s]   Draw the following graphs:
   (i)   the null graph N_5;
   (ii)   the complete graph K_6;
   (iii)   the complete bipartite graph $K_{2,4}$;
   (iv)   the union of $K_{1,3}$ and W_4;
   (v)   the complement of the cycle graph C_4.

3.2[s]   How many edges has each of the following graphs:
   (i) K_{10};   (ii) $K_{5,7}$;   (iii) Q_4;   (iv) W_8;   (v) the Petersen graph?

3.3   How many vertices and edges has each of the Platonic graphs?

3.4[s]   In the table of Fig. 2.9, locate all the regular graphs and the bipartite graphs.

3.5   Give an example (if it exists) of each of the following:
   (i)   a bipartite graph that is regular of degree 5;
   (ii)   a bipartite Platonic graph;
   (iii)   a complete graph that is a wheel;
   (iv)   a cubic graph with 11 vertices;
   (v)   a graph (other than K_4, $K_{4,4}$ or Q_4) that is regular of degree 4.

3.6[s]   Draw all the simple cubic graphs with at most 8 vertices.

3.7   The **complete tripartite graph** $K_{r,s,t}$ consists of three sets of vertices (of sizes r, s and t), with an edge joining two vertices if and only if they lie in different sets. Draw the graphs $K_{2,2,2}$ and $K_{3,3,2}$ and find the number of edges of $K_{3,4,5}$.

3.8   A simple graph that is isomorphic to its complement is **self-complementary**.
   (i)   Prove that, if G is self-complementary, then G has $4k$ or $4k+1$ vertices, where k is an integer.
   (ii)   Find all self-complementary graphs with 4 and 5 vertices.
   (iii)   Find a self-complementary graph with 8 vertices.

3.9*   The **line graph** $L(G)$ of a simple graph G is the graph whose vertices are in one–one correspondence with the *edges* of G, two vertices of $L(G)$ being adjacent if and only if the corresponding edges of G are adjacent.
   (i)   Show that K_3 and $K_{1,3}$ have the same line graph.
   (ii)   Show that the line graph of the tetrahedron graph is the octahedron graph.
   (iii)   Prove that, if G is regular of degree k, then $L(G)$ is regular of degree $2k-2$.
   (iv)   Find an expression for the number of edges of $L(G)$ in terms of the degrees of the vertices of G.
   (v)   Show that $L(K_5)$ is the complement of the Petersen graph.

3.10*   An **automorphism** φ of a simple graph G is a one–one mapping of the vertex set of G onto itself with the property that $\varphi(v)$ and $\varphi(w)$ are adjacent whenever v and w are. The **automorphism group** $\Gamma(G)$ of G is the group of automorphisms of G under composition.

(i)   Prove that the groups $\Gamma(G)$ and $\Gamma(\overline{G})$ are isomorphic.

(ii) Find the groups $\Gamma(K_n)$, $\Gamma(K_{r,s})$ and $\Gamma(C_n)$.

(iii) Use the results of parts (i) and (ii) and Exercise 3.9(v) to find the automorphism group of the Petersen graph.

4   Three puzzles

In this section we present three recreational puzzles that can be solved using ideas relating to graphs. In each puzzle, note how the use of a graph diagram makes the problem much easier to understand.

The eight circles problem

Place the letters A, B, C, D, E, F, G, H into the eight circles in Fig. 4.1, in such a way that no letter is adjacent to a letter that is next to it in the alphabet.

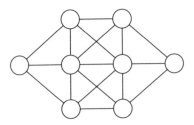

Fig. 4.1

First note that trying all the possibilities is not a practical proposition, as there are $8! = 40\ 320$ ways of placing eight letters into eight circles. We therefore need a more systematic approach.

Note that:

(i)   the easiest letters to place are A and H, because each has only one letter to which it cannot be adjacent (namely, B and G, respectively);

(ii) the hardest circles to fill are those in the middle, as each is adjacent to six others.

This suggests that we place A and H in the middle circles. If we place A to the left of H, then the only possible positions for B and G are as shown in Fig. 4.2.

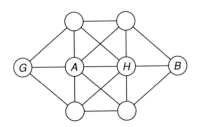

Fig. 4.2

The letter *C* must now be placed on the left-hand side of the diagram, and *F* must be placed on the right-hand side. It is then a simple matter to place the remaining letters, as shown in Fig. 4.3.

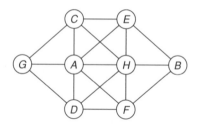

Fig. 4.3

Six people at a party

Show that, in any gathering of six people, there are either three people who all know each other or three people none of whom knows either of the other two.

To solve this, we draw a graph in which we represent each person by a vertex and join two vertices by a solid edge if the corresponding people know each other, and by a dotted edge if not. We must show that there is always a solid triangle or a dotted triangle.

Let *v* be any vertex. Then there must be exactly five edges incident with *v*, either solid or dashed, and so at least three of these edges must be of the same type. Let us assume that there are three solid edges (see Fig. 4.4); the case of at least three dashed edges is similar.

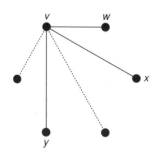

Fig. 4.4

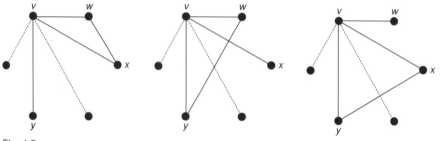

Fig. 4.5

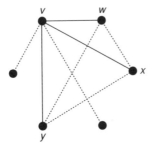

Fig. 4.6

If the people corresponding to the vertices w and x know each other, then v, w and x form a solid triangle, as required. Similarly, if the people corresponding to the vertices w and y, or to the vertices x and y, know each other, then we again obtain a solid triangle. These three cases are shown in Fig. 4.5.

Finally, if no two of the people corresponding to the vertices w, x and y know each other, then w, x and y form a dotted triangle, as required (see Fig. 4.6).

The four cubes problem

We conclude this section with a puzzle that has been popular under the name of *Instant Insanity*.

Given four cubes whose faces are coloured red, blue, green and yellow, as in Fig. 4.7, can we pile them up so that all four colours appear on each side of the resulting 4×1 stack?

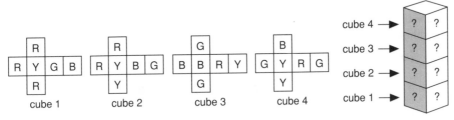

Fig. 4.7

Although these cubes can be stacked in thousands of different ways, there is essentially only one way that gives a solution.

We solve this problem by representing each cube by a graph with four vertices, R, B, G and Y, one for each colour. In each of these graphs, two vertices are adjacent if and only if the cube in question has the corresponding colours on *opposite* faces. The graphs corresponding to the cubes of Fig. 4.7 are shown in Fig. 4.8.

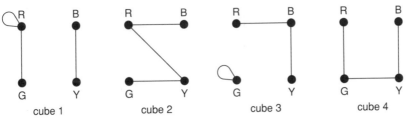

Fig. 4.8

We next superimpose these graphs to form a new graph G (see Fig. 4.9).

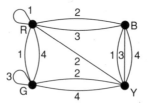

Fig. 4.9

A solution of the puzzle is obtained by finding two subgraphs H_1 and H_2 of G. The subgraph H_1 tells us which pair of colours appears on the front and back of each cube, and the subgraph H_2 tells us which pair of colours appears on the left and right. To this end, the subgraphs H_1 and H_2 have the following properties:

(a) *each subgraph contains exactly one edge from each cube*; this ensures that each cube has a front and back, and a left and right, and the subgraphs tell us which pairs of colours appear on these faces.

(b) *the subgraphs have no edges in common*; this ensures that the faces on the front and back are different from those on the sides.

(c) *each subgraph is regular of degree 2*; this tells us that each colour appears exactly twice on the sides of the stack (once on each side) and exactly twice on the front and back (once on the front and once on the back).

The subgraphs corresponding to our particular example are shown in Fig. 4.10, and the solution can be read from these subgraphs as in Fig. 4.11.

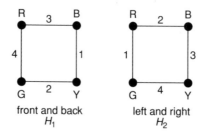

Fig. 4.10

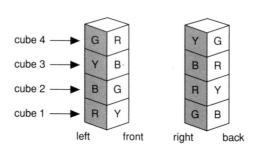

Fig. 4.11

Exercises 4

4.1 Find another solution of the eight circles problem.

4.2 Show that there is a gathering of five people in which there are no three people who all know each other and no three people none of whom knows either of the other two.

4.3 Find a solution of the four cubes problem for the set of cubes in Fig. 4.12.

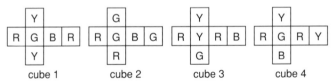

Fig. 4.12

4.4 Show that the four cubes problem in Fig. 4.13 has no solution.

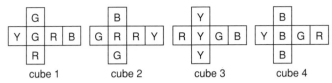

Fig. 4.13

4.5* Prove that the solution of the four cubes problem in the text is the only solution for that set of cubes.

Chapter 3

Paths and cycles

. . . So many paths that wind and wind,
While just the art of being kind
Is all the sad world needs.
Ella Wheeler Wilcox

Now that we have a reasonable armoury of graphs, we can look at their properties. To do this, we need some definitions that describe ways of 'going from one vertex to another'. We give these definitions in Section 5 and prove some results on connectivity. In Sections 6 and 7 we study two particular types of graphs, those with trails containing every edge, and those with cycles containing every vertex. We conclude this chapter, in Section 8, with some applications of paths and cycles.

5 Connectivity

Given a graph G, a **walk** in G is a finite sequence of edges of the form $v_0v_1, v_1v_2, \ldots,$ $v_{m-1}v_m$, also denoted by $v_0 \to v_1 \to v_2 \to \cdots \to v_m$, in which any two consecutive edges are adjacent or identical. Such a walk determines a sequence of vertices $v_0, v_1,$ $\ldots, v_m$. We call v_0 the **initial vertex** and v_m the **final vertex** of the walk, and speak of a **walk from v_0 to v_m**. The number of edges in a walk is called its **length**; for example, in Fig. 5.1, $v \to w \to x \to y \to z \to z \to y \to w$ is a walk of length 7 from v to w.

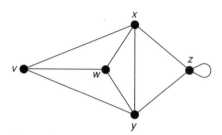

Fig. 5.1

The concept of a walk is usually too general for our purposes, so we impose some restrictions. A walk in which all the edges are distinct is a **trail**. If, in addition, the vertices $v_0, v_1, \ldots, v_m$ are distinct (except, possibly, $v_0 = v_m$), then the trail is a **path**. A path or trail is **closed** if $v_0 = v_m$, and a closed path containing at least one edge is a **cycle**. Note that any loop or pair of multiple edges is a cycle.

To clarify these concepts, consider Fig. 5.1. We see that $v \to w \to x \to y \to z \to z \to x$ is a trail, $v \to w \to x \to y \to z$ is a path, $v \to w \to x \to y \to z \to x \to v$ is a closed trail, and $v \to w \to x \to y \to v$ is a cycle. A cycle of length 3, such as $v \to w \to x \to v$, is a **triangle**.

Note that a graph is **connected** if and only if there is a path between each pair of vertices (see Fig. 5.2).

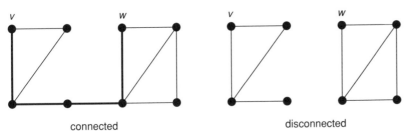

connected   disconnected

Fig. 5.2

Note also that G is a bipartite graph if and only if each cycle of G has even length. We prove one half of this result here, leaving the proof of its converse to you (see Exercise 5.3).

> **THEOREM 5.1.** *If G is a bipartite graph, then each cycle of G has even length.*

Proof. Since G is bipartite, we can split its vertex set into two disjoint sets A and B so that each edge of G joins a vertex of A and a vertex of B. Let $v_0 \to v_1 \to \cdots \to v_m \to v_0$ be a cycle in G, and assume (without loss of generality) that v_0 is in A. Then v_1 is in B, v_2 is in A, and so on. Since v_m must be in B, the cycle has even length. //

We now investigate bounds for the number of edges of a simple connected graph on n vertices. Such a graph has fewest edges when it has no cycles, and most edges when it is a complete graph. This implies that the number of edges must lie between $n - 1$ and $n(n - 1)/2$. In fact, we prove a stronger theorem that includes this result as a special case.

> **THEOREM 5.2.** *Let G be a simple graph on n vertices. If G has k components, then the number m of edges of G satisfies*
> $$n - k \le m \le (n - k)(n - k + 1)/2.$$

Proof. We prove the lower bound $m \ge n - k$ by induction on the number of edges of G, the result being trivial if G is a null graph. If G contains as few edges as possible (say m_0), then the removal of any edge of G must increase the number of components by 1, and the graph that remains has n vertices, $k + 1$ components, and $m_0 - 1$ edges. It follows from the induction hypothesis that $m_0 - 1 \ge n - (k + 1)$, giving $m_0 \ge n - k$, as required.

To prove the upper bound, we can assume that each component of G is a complete graph. Suppose, then, that there are two components C_i and C_j with n_i and n_j vertices, respectively, where $n_i \geq n_j > 1$. If we replace C_i and C_j by complete graphs on $n_i + 1$ and $n_j - 1$ vertices, then the total number of vertices remains unchanged, and the number of edges is changed by

$$\{(n_i + 1)n_i - n_i(n_i - 1)\}/2 - \{n_j(n_j - 1) - (n_j - 1)(n_j - 2)\}/2 = n_i - n_j + 1,$$

which is positive. It follows that, in order to attain the maximum number of edges, G must consist of a complete graph on $n - k + 1$ vertices and $k - 1$ isolated vertices. The result now follows. //

COROLLARY 5.3. *Any simple graph with n vertices and more than $(n-1)(n-2)/2$ edges is connected.*

Another approach used in the study of connected graphs is to ask 'how connected is a connected graph?' One interpretation of this is to ask how many edges or vertices must be removed in order to disconnect the graph. We conclude this section with some terms that are useful when discussing such questions.

A **disconnecting set** in a connected graph G is a set of edges whose removal disconnects G. For example, in the graph of Fig. 5.3, the sets $\{e_1, e_2, e_5\}$ and $\{e_3, e_6, e_7, e_8\}$ are both disconnecting sets of G; the disconnected graph left after removal of the second is shown in Fig. 5.4.

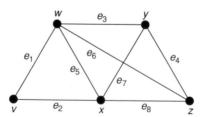

Fig. 5.3

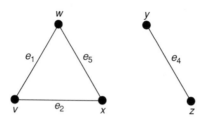

Fig. 5.4

We further define a **cutset** to be a disconnecting set, no proper subset of which is a disconnecting set. In the above example, only the second disconnecting set is a cutset. Note that the removal of the edges in a cutset always leaves a graph with exactly two components. If a cutset has only one edge e, we call e a **bridge** (see Fig. 5.5).

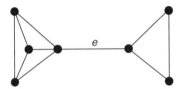

Fig. 5.5

These definitions can easily be extended to disconnected graphs. If G is any such graph, a **disconnecting set** of G is a set of edges whose removal increases the number of components of G, and a **cutset** of G is a disconnecting set, no proper subset of which is a disconnecting set.

If G is connected, its **edge connectivity** $\lambda(G)$ is the size of the smallest cutset in G. Thus $\lambda(G)$ is the minimum number of edges that we need to delete in order to disconnect G. For example, if G is the graph of Fig. 5.3, then $\lambda(G) = 2$, corresponding to the cutset $\{e_1, e_2\}$. We also say that G is **k-edge connected** if $\lambda(G) \geq k$. Thus the graph of Fig. 5.3 is 1-edge connected and 2-edge connected, but not 3-edge connected.

We also need the analogous concepts for the removal of vertices. A **separating set** in a connected graph G is a set of vertices whose deletion disconnects G; recall that when we delete a vertex, we also remove its incident edges. For example, in the graph of Fig. 5.3, the sets $\{w, x\}$ and $\{w, x, y\}$ are separating sets of G; the disconnected graph left after removal of the first is shown in Fig. 5.6. If a separating set contains only one vertex v, we call v a **cut-vertex** (see Fig. 5.7). These definitions extend immediately to disconnected graphs, as above.

Fig. 5.6

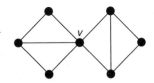

Fig. 5.7

If G is connected and not a complete graph, its (vertex) **connectivity** $\kappa(G)$ is the size of the smallest separating set in G. Thus $\kappa(G)$ is the minimum number of vertices that we need to delete in order to disconnect G. For example, if G is the graph of Fig. 5.3, then $\kappa(G) = 2$, corresponding to the separating set $\{w, x\}$. We also say that G is **k-connected** if $\kappa(G) \geq k$. Thus the graph of Fig. 5.3 is 1-connected and 2-connected, but not 3-connected. It can be proved that, if G is any connected graph, then $\kappa(G) \leq \lambda(G)$.

Finally, we remark that there are striking and unexpected similarities between the properties of cycles and cutsets. To see this, look at Exercises 5.11, 5.12, 5.13, 6.8 and 9.10. The reason for this is revealed in Chapter {9}, when all will become clear!

Exercises 5

5.1[s]   In the Petersen graph, find
  (i)   a trail of length 5;
  (ii)   a path of length 9;
  (iii) cycles of lengths 5, 6, 8 and 9;
  (iv)   cutsets with 3, 4 and 5 edges.

5.2[s]   The **girth** of a graph is the length of its shortest cycle. Write down the girths of (i) K_9; (ii) $K_{5,7}$; (iii) C_8; (iv) W_8; (v) Q_5; (vi) the Petersen graph; (vii) the graph of the dodecahedron.

5.3   Prove the converse of Theorem 5.1 – that if each cycle of a graph G has even length, then G is bipartite.

5.4[s]   Prove that a simple graph and its complement cannot both be disconnected.

5.5[s]   Write down $\kappa(G)$ and $\lambda(G)$ for each of the following graphs G:
(i) C_6; (ii) W_6; (iii) $K_{4,7}$; (iv) Q_4.

5.6   (i)   Show that, if G is a connected graph with minimum degree k, then $\lambda(G) \le k$.
  (ii) Draw a graph G with minimum degree k for which $\kappa(G) < \lambda(G) < k$.

5.7   (i)   Prove that a graph is 2-connected if and only if each pair of vertices are contained in a common cycle.
  (ii) Write down a corresponding statement for a 2-edge-connected graph.

5.8   Let G be a connected graph with vertex set $\{v_1, v_2, \ldots, v_n\}$, m edges and t triangles.
  (i)   Given that $\mathbf{A}$ is the adjacency matrix of G, prove that the number of walks of length 2 from v_i to v_j is the ij-th entry of the matrix $\mathbf{A}^2$.
  (ii)   Deduce that $2m =$ the sum of the diagonal entries of $\mathbf{A}^2$.
  (iii) Obtain a result for the number of walks of length 3 from v_i to v_j, and deduce that $6t =$ the sum of the diagonal entries of $\mathbf{A}^3$.

5.9   In a connected graph, the **distance** $d(v,w)$ from v to w is the length of the shortest path from v to w.
  (i)   If $d(v, w) \ge 2$, show that there exists a vertex z such that $d(v, z) + d(z, w) = d(v, w)$.
  (ii)   In the Petersen graph, show that $d(v, w) = 1$ or 2, for any distinct vertices v and w.

5.10*   Let G be a simple graph on $2k$ vertices containing no triangles. Show, by induction on k, that G has at most k^2 edges, and give an example of a graph for which this upper bound is achieved. (This result is often called *Turán's extremal theorem*.)

5.11*   (i)   Prove that, if two distinct cycles of a graph G each contain an edge e, then G has a cycle that does not contain e.
  (ii) Prove a similar result with 'cycle' replaced by 'cutset'.

5.12*   (i)   Prove that, if C is a cycle and C^* is a cutset of a connected graph G, then C and C^* have an even number of edges in common.
  (ii) Prove that, if S is any set of edges of G with an even number of edges in common with each cutset of G, then S can be split into edge-disjoint cycles.

5.13*   A set E of edges of a graph G is **independent** if E contains no cycle of G. Prove that:

(i)   any subset of an independent set is independent;

(ii) if *I* and *J* are independent sets of edges with |*J*| > |*I*|, then there is an edge *e* that lies in *J* but not in *I* with the property that *I* ∪ {*e*} is independent.

Show also that (i) and (ii) still hold if we replace the word 'cycle' by cutset'.

6 Eulerian graphs

A connected graph *G* is **Eulerian** if there exists a closed trail containing every edge of *G*. Such a trail is an **Eulerian trail**. Note that this definition requires each edge to be traversed once and once only. A non-Eulerian graph *G* is **semi-Eulerian** if there exists a trail containing every edge of *G*. Figs 6.1, 6.2 and 6.3 show graphs that are Eulerian, semi-Eulerian and non-Eulerian, respectively.

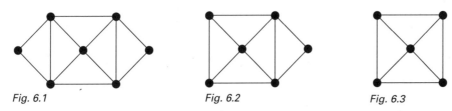

Fig. 6.1   Fig. 6.2   Fig. 6.3

Problems on Eulerian graphs frequently appear in books on recreational mathematics. A typical problem might ask whether a given diagram can be drawn without lifting one's pencil from the paper and without repeating any lines. The name 'Eulerian' arises from the fact that Euler was the first person to solve the famous **Königsberg bridges problem** which asks whether you can cross each of the seven bridges in Fig. 6.4 exactly once and return to your starting point. This is equivalent to asking whether the graph in Fig. 6.5 has an Eulerian trail. A translation of Euler's paper, and a discussion of various related topics, may be found in Biggs, Lloyd and Wilson [11].

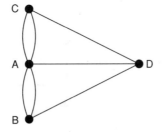

Fig. 6.4

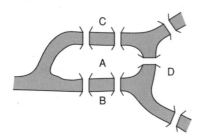

Fig. 6.5

One question that immediately arises is 'can one find necessary and sufficient conditions for a graph to be Eulerian?' Before answering this question in Theorem 6.2, we prove a simple lemma.

> **LEMMA 6.1.** *If G is a graph in which the degree of each vertex is at least 2, then G contains a cycle.*

Proof. If G has any loops or multiple edges, the result is trivial. We can therefore suppose that G is a simple graph. Let v be any vertex of G. We construct a walk $v \rightarrow v_1 \rightarrow v_2 \rightarrow \cdots$ inductively by choosing v_1 to be any vertex adjacent to v and, for each $i > 1$, choosing v_{i+1} to be any vertex adjacent to v_i except v_{i-1}; the existence of such a vertex is guaranteed by our hypothesis. Since G has only finitely many vertices, we must eventually choose a vertex that has been chosen before. If v_k is the first such vertex, then that part of the walk lying between the two occurrences of v_k is the required cycle. //

> **THEOREM 6.2 (Euler 1736).** *A connected graph G is Eulerian if and only if the degree of each vertex of G is even.*

Proof. $\Rightarrow$ Suppose that P is an Eulerian trail of G. Whenever P passes through a vertex, there is a contribution of 2 towards the degree of that vertex. Since each edge occurs exactly once in P, each vertex must have even degree.

$\Leftarrow$ The proof is by induction on the number of edges of G. Suppose that the degree of each vertex is even. Since G is connected, each vertex has degree at least 2 and so, by Lemma 6.1, G contains a cycle C. If C contains every edge of G, the proof is complete. If not, we remove from G the edges of C to form a new, possibly disconnected, graph H with fewer edges than G and in which each vertex still has even degree. By the induction hypothesis, each component of H has an Eulerian trail. Since each component of H has at least one vertex in common with C, by connectedness, we obtain the required Eulerian trail of G by following the edges of C until a non-isolated vertex of H is reached, tracing the Eulerian trail of the component of H that contains that vertex, and then continuing along the edges of C until we reach a vertex belonging to another component of H, and so on. The whole process terminates when we return to the initial vertex (see Fig. 6.6). //

This proof can easily be modified to prove the following two results. We omit the details.

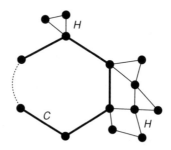

Fig. 6.6

COROLLARY 6.3. *A connected graph is Eulerian if and only if its set of edges can be split up into disjoint cycles.*

COROLLARY 6.4. *A connected graph is semi-Eulerian if and only if it has exactly two vertices of odd degree.*

Note that, in a semi-Eulerian graph, any semi-Eulerian trail must have one vertex of odd degree as its initial vertex and the other as its final vertex. Note also that, by the handshaking lemma, a graph cannot have exactly one vertex of odd degree.

We conclude our discussion of Eulerian graphs with an algorithm for constructing an Eulerian trail in a given Eulerian graph. The method is known as **Fleury's algorithm**.

THEOREM 6.5. *Let G be an Eulerian graph. Then the following construction is always possible, and produces an Eulerian trail of G.*

Start at any vertex u and traverse the edges in an arbitrary manner, subject only to the following rules:

(i) erase the edges as they are traversed, and if any isolated vertices result, erase them too;

(ii) at each stage, use a bridge only if there is no alternative.

Proof. We show first that the construction can be carried out at each stage. Suppose that we have just reached a vertex v. If $v \neq u$, then the subgraph H that remains is connected and contains only two vertices of odd degree, u and v. To show that the construction can be carried out, we must show that the removal of the next edge does not disconnect H – or, equivalently, that v is incident with at most one bridge. But if this is not the case, then there exists a bridge vw such that the component K of $H - vw$ containing w does not contain u (see Fig. 6.7). Since the vertex w has odd degree in K, some other vertex of K must also have odd degree, giving the required contradiction. If $v = u$, the proof is almost identical, as long as there are still edges incident with u.

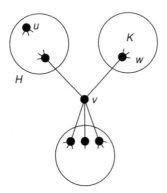

Fig. 6.7

It remains only to show that this construction always yields a complete Eulerian trail. But this is clear, since there can be no edges of G remaining untraversed when the last edge incident to *u* is used, since otherwise the removal of some *earlier* edge adjacent to one of these edges would have disconnected the graph, contradicting (ii). //

Exercises 6

6.1ˢ   Which of the following graphs are Eulerian? semi-Eulerian?
    (i)   the complete graph K_5;
    (ii) the complete bipartite graph $K_{2,3}$;
    (iii) the graph of the cube;
    (iv) the graph of the octahedron;
    (v) the Petersen graph.

6.2ˢ   In the table of Fig. 2.9, locate all the Eulerian and semi-Eulerian graphs.

6.3   (i)   For which values of *n* is K_n Eulerian?
    (ii) Which complete bipartite graphs are Eulerian?
    (iii) Which Platonic graphs are Eulerian?
    (iv) For which values of *n* is the wheel W_n Eulerian?
    (v) For which values of *k* is the *k*-cube Q_k Eulerian?

6.4ˢ   Let G be a connected graph with *k* (> 0) vertices of odd degree.
    (i) Show that the minimum number of trails, that together include every edge of G and that have no edges in common, is *k*/2.
    (ii) How many continuous pen-strokes are needed to draw the diagram in Fig. 6.8 without repeating any line?

Fig. 6.8

6.5ˢ   Use Fleury's algorithm to produce an Eulerian trail for the graph in Fig. 6.9.

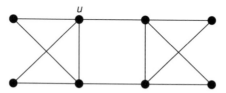

Fig. 6.9

6.6   (i) Show that the line graph of a simple Eulerian graph is Eulerian.
    (ii) If the line graph of a simple graph G is Eulerian, must G be Eulerian?

6.7   An Eulerian graph is **randomly traceable** from a vertex *v* if, whenever we start from *v* and traverse the graph in an arbitrary way never using any edge twice, we eventually obtain an Eulerian trail.
    (i) Show that the graph in Fig. 6.10 is randomly traceable.

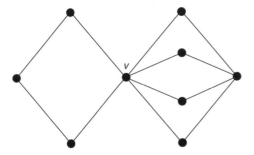

Fig. 6.10

    (ii)   Give an example of an Eulerian graph that is not randomly traceable.

    (iii) Why might a randomly traceable graph be suitable for the layout of an exhibition?

6.8*   Let V be the vector space associated with a graph G (see Exercise 2.15).

    (i)   Use Corollary 6.3 to show that, if C and D are cycles of G, then their sum $C + D$ can be written as a union of edge-disjoint cycles.

    (ii) Deduce that the set of such unions of cycles of C forms a subspace W of V (the **cycle subspace** of G), and find its dimension.

    (iii) Show that the set of unions of edge-disjoint cutsets of G forms a subspace W^* of V (the **cutset subspace** of G), and find its dimension.

7 Hamiltonian graphs

In the previous section we discussed whether there exists a closed trail that includes every edge of a given connected graph G. A similar problem is to determine whether there exists a closed trail passing exactly once through each vertex of G. Note that such a trail must be a cycle, except when G is the graph N_1. Such a cycle is a **Hamiltonian cycle** and G is a **Hamiltonian graph**. A non-Hamiltonian graph G is semi-Hamiltonian if there exists a path passing through every vertex. Figs 7.1, 7.2 and 7.3 show graphs that are Hamiltonian, semi-Hamiltonian and non-Hamiltonian, respectively.

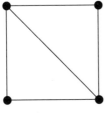

Fig. 7.1

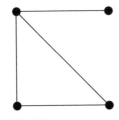

Fig. 7.2

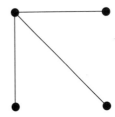

Fig. 7.3

The name 'Hamiltonian cycle' arises from the fact that Sir William Hamilton investigated their existence in the dodecahedron graph, although a more general problem had been studied earlier by the Rev. T.P. Kirkman. Such a cycle is shown in Fig. 7.4, with heavy lines denoting its edges.

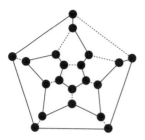

Fig. 7.4

In Theorem 6.2 and Corollary 6.3 we obtained necessary and sufficient conditions for a connected graph to be Eulerian, and we may hope to obtain similar characterizations for Hamiltonian graphs. As it happens, the finding of such a characterization is one of the major unsolved problems of graph theory! In fact, little is known in general about Hamiltonian graphs. Most existing theorems have the form, 'if G has enough edges, then G is Hamiltonian'. Probably the most celebrated of these is due to G.A. Dirac, and known as **Dirac's theorem**. We deduce it from the following more general result of O. Ore.

> **THEOREM 7.1 (Ore, 1960).** *If G is a simple graph with n (≥ 3) vertices, and if*
>
> $$\deg(v) + \deg(w) \geq n$$
>
> *for each pair of non-adjacent vertices v and w, then G is Hamiltonian.*

Proof. We assume the theorem false, and derive a contradiction. So let G be a *non-Hamiltonian* graph with n vertices, satisfying the given condition on the vertex degrees. By adding extra edges if necessary, we may assume that G is 'only just' non-Hamiltonian, in the sense that the addition of any further edge gives a Hamiltonian graph. (Note that adding an extra edge does not violate the condition on the vertex degrees.) It follows that G contains a path $v_1 \rightarrow v_2 \rightarrow \cdots \rightarrow v_n$ passing through every vertex. But since G is non-Hamiltonian, the vertices v_1 and v_n are not adjacent, and so $\deg(v_1) + \deg(v_n) \geq n$. It follows that there must be some vertex v_i adjacent to v_1 with the property that v_{i-1} is adjacent to v_n (see Fig. 7.5). But this gives us the required contradiction since

$$v_1 \rightarrow v_2 \rightarrow \cdots \rightarrow v_{i-1} \rightarrow v_n \rightarrow v_{n-1} \rightarrow \cdots \rightarrow v_{i+1} \rightarrow v_i \rightarrow v_1$$

is then a Hamiltonian cycle. //

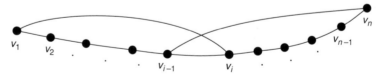

Fig. 7.5

> **COROLLARY 7.2 (Dirac, 1952).** *If G is a simple graph with n (≥ 3) vertices, and if $\deg(v) > n/2$ for each vertex v, then G is Hamiltonian.*

Proof. The result follows immediately from Theorem 7.1, since deg(v)+deg(w) $\geq n$ for each pair of vertices v and w, whether adjacent or not. //

Exercises 7

7.1^s   Which of the following graphs are Hamiltonian? semi-Hamiltonian?
    (i)   the complete graph K_5;
    (ii) the complete bipartite graph $K_{2,3}$;
    (iii) the graph of the octahedron;
    (iv) the wheel W_6;
    (v) the 4-cube Q_4.

7.2^s   In the table of Fig. 2.9, locate all the Hamiltonian and semi-Hamiltonian graphs.

7.3   (i)   For which values of n is K_n Hamiltonian?
    (ii) Which complete bipartite graphs are Hamiltonian?
    (iii) Which Platonic graphs are Hamiltonian?
    (iv) For which values of n is the wheel W_n Hamiltonian?
    (v) For which values of k is the k-cube Q_k Hamiltonian?

7.4   Show that the Grötzsch graph in Fig. 7.6 is Hamiltonian.

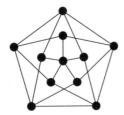

Fig. 7.6

7.5   (i)   Prove that, if G is a bipartite graph with an odd number of vertices, then G is non-Hamiltonian.
    (ii) Deduce that the graph in Fig. 7.7 is non-Hamiltonian.

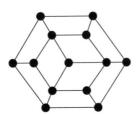

Fig. 7.7

    (iii) Show that, if n is odd, it is not possible for a knight to visit all the squares of an $n \times n$ chessboard exactly once by knight's moves and return to its starting point.

7.6^s   Give an example to show that the condition 'deg(v) $\geq n/2$', in the statement of Dirac's theorem, cannot be replaced by 'deg(v) $\geq (n-1)/2$'.

7.7   (i)   Let G be a graph with n vertices and $[(n-1)(n-2)/2] + 2$ edges. Use Theorem 7.1 to prove that G is Hamiltonian.
    (ii) Find a non-Hamiltonian graph with n vertices and $[(n-1)(n-2)/2] + 1$ edges.

7.8*   Prove that the Petersen graph is non-Hamiltonian.

7.9*   Let G be a Hamiltonian graph and let S be any set of k vertices in G. Prove that the graph $G - S$ has at most k components.

7.10*   (i)   Find four Hamiltonian cycles in K_9, no two of which have an edge in common.
(ii)   What is the maximum number of edge-disjoint Hamiltonian cycles in K_{2k+1}?

8   Some algorithms

Most important advances in graph theory arose as a result of attempts to solve particular practical problems – Euler and the bridges of Königsberg (Section 6), Cayley and the enumeration of chemical molecules (Section 11), and Kirchhoff's work on electrical networks (Section 11), to name but three. Much present-day interest in the subject is due to the fact that, quite apart from being an elegant mathematical discipline in its own right, graph theory is applied in a wide range of areas (see the Preface). We cannot discuss a large number of these applications in a book of this size. You should consult Berge [6], Bondy and Murty [7], Deo [13], Tucker [20] and Wilson and Beineke [21] for a wide range of practical problems, often with algorithms or flow-charts for their solution.

In this section we briefly describe three problems that relate to the present chapter – the *shortest path problem*, the *Chinese postman problem* and the *travelling salesman problem*. The first of these can be solved by an efficient **algorithm** – that is, a finite step-by-step procedure that quickly gives the solution. The second problem can also be solved by an algorithm, but we consider only a special case here. For the third problem, no efficient algorithms are known; we must therefore choose between algorithms that take a long time to carry out and heuristic algorithms that are quick to apply but give only an approximation to the solution.

The shortest path problem
Suppose that we have a 'map' of the form shown in Fig. 8.1, in which the letters *A*–*L* refer to towns that are connected by roads. If the lengths of these roads are as marked, what is the length of the shortest path from A to L?

Note that the numbers in the diagram need not refer to the lengths of the roads, but could refer to the times taken to travel along them, or to the costs of doing so. Thus, if

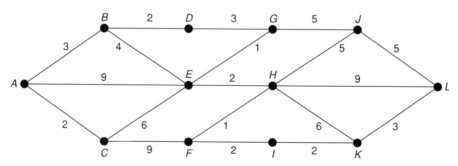

Fig. 8.1

we have an algorithm for solving this problem in its original formulation, then this algorithm can also be used to give the quickest or cheapest route.

Note that an upper bound for the answer can easily be obtained by taking *any* path from A to L and calculating its length. For example, the path $A \to B \to D \to G \to J \to L$ has total length 18, and so the length of the shortest path cannot exceed 18.

In such problems our 'map' can be regarded as a connected graph in which a non-negative number is assigned to each edge. Such a graph is called a **weighted graph**, and the number assigned to each edge e is the **weight** of e, denoted by $w(e)$. The problem is to find a path from A to L with minimum total weight. Note that, if we have a weighted graph in which each edge has weight 1, then the problem reduces to that of finding the number of edges in the shortest path from A to L.

There are several methods that can be used to solve this problem. One way is to make a model of the map by knotting together pieces of string whose lengths are proportional to the lengths of the roads. To find the shortest path, take hold of the knots corresponding to A and L – and pull tight!

However, there is a more mathematical way of approaching this problem. The idea is to move across the graph from left to right, associating with each vertex V a number $l(V)$ indicating the shortest distance from A to V. This means that, when we reach a vertex such as K in Fig. 8.1, then $l(K)$ is either $l(H) + 6$ or $l(I) + 2$, whichever is the smaller.

To apply the algorithm, we first assign A the label 0 and give B, E and C the temporary labels $l(A) + 3$, $l(A) + 9$ and $l(A) + 2$ – that is, 3, 9 and 2. We take the *smallest* of these, and write $l(C) = 2$. C is now permanently labelled 2.

We next look at the vertices adjacent to C. We assign F the temporary label $l(C) + 9 = 11$, and we can lower the temporary label at E to $l(C) + 6 = 8$. The smallest temporary label is now 3 (at B), so we write $l(B) = 3$. B is now permanently labelled 3.

Now we look at the vertices adjacent to B. We assign D the temporary label $l(B) + 2 = 5$, and we can lower the temporary label at E to $l(B) + 4 = 7$. The smallest temporary label is now 5 (at D), so we write $l(D) = 5$. D is now permanently labelled 5.

Continuing in this way, we successively obtain the permanent labels $l(E) = 7$, $l(G) = 8$, $l(H) = 9$, $l(F) = 10$, $l(I) = 12$, $l(J) = 13$, $l(K) = 14$, $l(L) = 17$. It follows that the shortest path from A to L has length 17. It is shown in Fig. 8.2, with circled numbers representing the labels at the vertices.

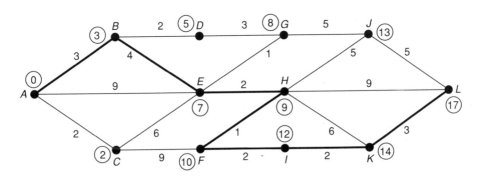

Fig. 8.2

In Section 22 we see how this algorithm can be adapted to yield the *longest* path in a digraph, and we illustrate its use in critical path analysis.

The Chinese postman problem

In this problem, discussed by the Chinese mathematician Mei-Ku Kwan, a postman wishes to deliver his letters, covering the least possible total distance and returning to his starting point. He must obviously traverse each road in his route at least once, but should avoid covering too many roads more than once.

This problem can be reformulated in terms of a weighted graph, where the graph corresponds to the network of roads, and the weight of each edge is the length of the corresponding road. In this reformulation, the requirement is to find a closed walk of minimum total weight that includes each edge at least once. If the graph is Eulerian, then any Eulerian trail is a closed walk of the required type. Such an Eulerian trail can be found by Fleury's algorithm (see Section 6). If the graph is not Eulerian, then the problem is much harder, although an efficient algorithm for its solution is known. To illustrate the ideas involved, we look at a special case, in which exactly two vertices have odd degree (see Fig. 8.3).

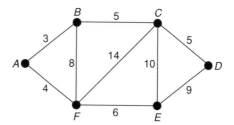

Fig. 8.3

Since vertices B and E are the only vertices of odd degree, we can find a semi-Eulerian trail from B to E covering each edge exactly once. In order to return to the starting point, covering the least possible distance, we now find the shortest path from E to B using the algorithm described above. The solution of the Chinese postman problem is then obtained by taking this shortest path $E \rightarrow F \rightarrow A \rightarrow B$, together with the original semi-Eulerian trail, giving a total distance of $13 + 64 = 77$. Note that, if we combine the shortest path and the semi-Eulerian trail, we get an Eulerian graph (see Fig. 8.4).

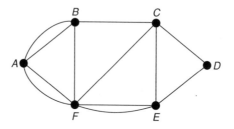

Fig. 8.4

A fuller discussion of the Chinese postman problem can be found in Bondy and Murty [7].

The travelling salesman problem

In this problem, a travelling salesman wishes to visit several given cities and return to his starting point, covering the least possible total distance. For example, if there are five cities A, B, C, D and E, and if the distances are as given in Fig. 8.5, then the shortest possible route is $A \rightarrow B \rightarrow D \rightarrow E \rightarrow C \rightarrow A$, giving a total distance of 26, as can be seen by inspection.

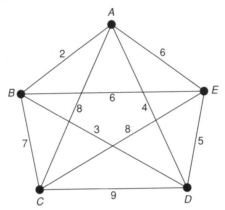

Fig. 8.5

This problem can also be reformulated in terms of weighted graphs. In this case, the requirement is to find a Hamiltonian cycle of least possible total weight in a weighted complete graph. Note that, as in the shortest path problem, the numbers can also refer to the times taken to travel between the cities, or the costs involved in doing so. Thus, if we could find an efficient algorithm for solving the travelling salesman problem in its original formulation, then we could apply the same algorithm to find the quickest or the cheapest route.

One possible algorithm is to calculate the total distance for *all* possible Hamiltonian cycles, but this is far too complicated for more than about five cities. For example, if there are 20 cities, then the number of possible cycles is (19!)/2, which is about 6×10^{16}. Various other algorithms have been proposed, but they take too long to apply. On the other hand, there are several heuristic algorithms that quickly tell us *approximately* what the shortest distance is. One of these procedures is described in Section 11.

Exercises 8

8.1[s]   Use the shortest path algorithm to find a shortest path from A to G in the weighted graph of Fig. 8.6.

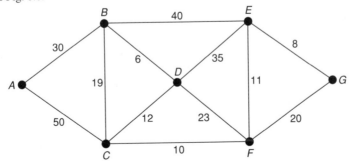

Fig. 8.6

8.2   Use the shortest path algorithm to find the shortest path from *L* to *A* in Fig. 8.1, and check that your answer agrees with that given in Fig. 8.2.

8.3   Show how the shortest path algorithm can be adapted to yield the *longest* path from *A* to *L* in Fig. 8.1.

8.4*   Find the shortest path from *S* to each other vertex in the weighted graph of Fig. 8.7.

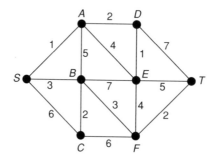

Fig. 8.7

8.5s   Solve the Chinese postman problem for the weighted graph of Fig. 8.8.

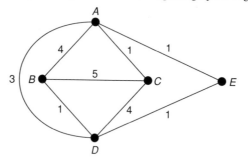

Fig. 8.8

8.6s   Solve the travelling salesman problem for the weighted graph of Fig. 8.9.

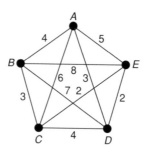

Fig. 8.9

8.7   Find the Hamiltonian cycle of *greatest* weight in the graph of Fig. 8.5.

Chapter 4

Trees

A fool sees not the same tree that a wise man sees.
William Blake

We are all familiar with the idea of a family tree. In this chapter, we study trees in general, with special reference to spanning trees in a connected graph and to Cayley's celebrated result on the enumeration of labelled trees. The chapter concludes with some further applications.

9   Properties of trees

A **forest** is a graph that contains no cycles, and a connected forest is a **tree**. For example, Fig. 9.1 shows a forest with four components, each of which is a tree[†]. Note that trees and forests are simple graphs.

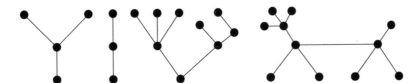

Fig. 9.1

In many ways a tree is the simplest non-trivial type of graph. As we shall see, it has several 'nice' properties, such as the fact that any two vertices are connected by a unique path. In trying to prove a general result for graphs, it is sometimes convenient to start by trying to prove it for trees. In fact, several conjectures that have not been proved for arbitrary graphs are known to be true for trees.

The following theorem lists some simple properties of trees.

[†] The last tree in Fig. 9.1 is particularly well known for its bark.

THEOREM 9.1. *Let T be a graph with n vertices. Then the following statements are equivalent:*

   (i)   *T is a tree;*
  (ii) *T contains no cycles, and has n–1 edges;*
 (iii) *T is connected, and has n–1 edges;*
 (iv) *T is connected, and each edge is a bridge;*
  (v) *any two vertices of T are connected by exactly one path;*
 (vi) *T contains no cycles, but the addition of any new edge creates exactly one cycle.*

Proof. If $n = 1$, all six results are trivial; we therefore assume that $n \geq 2$.

(i) $\Rightarrow$ (ii). Since T contains no cycles, the removal of any edge must disconnect T into two graphs, each of which is a tree. It follows by induction that the number of edges in each of these two trees is one fewer than the number of vertices. We deduce that the total number of edges of T is $n - 1$.

(ii) $\Rightarrow$ (iii). If T is disconnected, then each component of T is a connected graph with no cycles and hence, by the previous part, the number of vertices in each component exceeds the number of edges by 1. It follows that the total number of vertices of T exceeds the total number of edges by at least 2, contradicting the fact that T has $n - 1$ edges.

(iii) $\Rightarrow$ (iv). The removal of any edge results in a graph with n vertices and $n - 2$ edges, which must be disconnected by Theorem 5.2.

(iv) $\Rightarrow$ (v). Since T is connected, each pair of vertices is connected by at least one path. If a given pair of vertices is connected by two paths, then they enclose a cycle, contradicting the fact that each edge is a bridge.

(v) $\Rightarrow$ (vi). If T contained a cycle, then any two vertices in the cycle would be connected by at least two paths, contradicting statement (v). If an edge e is added to T, then, since the vertices incident with e are already connected in T, a cycle is created. The fact that only one cycle is obtained follows from Exercise 5.11.

(vi) $\Rightarrow$ (i). Suppose that T is disconnected. If we add to T any edge joining a vertex of one component to a vertex in another, then no cycle is created. //

COROLLARY 9.2. *If G is a forest with n vertices and k components, then G has n – k edges.*

Proof. Apply Theorem 9.1(iii) above to each component of G. //

Note that, by the handshaking lemma, the sum of the degrees of the n vertices of a tree is equal to twice the number of edges $(= 2n - 2)$. It follows that *if $n > 2$, any tree on n vertices has at least two end-vertices.*

Given any connected graph G, we can choose a cycle and remove any one of its edges, and the resulting graph remains connected. We repeat this procedure with one of the remaining cycles, continuing until there are no cycles left. The graph that remains is a tree that connects all the vertices of G. It is called a **spanning tree** of G. An example of a graph and one of its spanning trees appears in Fig 9.2.

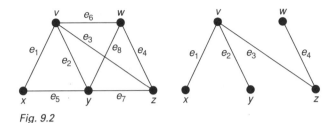

Fig. 9.2

More generally, if G is an arbitrary graph with n vertices, m edges and k components, then we can carry out this procedure on each component of G. The result is called a **spanning forest**, and the total number of edges removed in this process is the **cycle rank** of G, denoted by $\gamma(G)$. Note that $\gamma(G) = m - n + k$, which is a non-negative integer by Theorem 5.2. It is convenient also to define the **cutset rank** of G to be the number of edges in a spanning forest, denoted by $\xi(G)$. Note that $\xi(G) = n - k$. Some properties of the cutset rank are given in Exercise 9.12.

Before proceeding, we prove a couple of simple results concerning spanning forests. In this theorem, the **complement** of a spanning forest T of a (not necessarily simple) graph G is the graph obtained from G by removing the edges of T.

> **THEOREM 9.3.** *If T is any spanning forest of a graph G, then*
> *(i)   each cutset of G has an edge in common with T;*
> *(ii) each cycle of G has an edge in common with the complement of T.*

Proof.   (i) Let C^* be a cutset of G, the removal of which splits a component of G into two subgraphs H and K. Since T is a spanning forest, T must contain an edge joining a vertex of H to a vertex of K, and this edge is the required edge.

(ii) Let C be a cycle of G having no edge in common with the complement of T. Then C must be contained in T, which is a contradiction. //

Closely linked with the idea of a spanning forest T of a graph G is that of the fundamental set of cycles associated with T. This is formed as follows: if we add to T any edge of G not contained in T, then by Theorem 9.1(vi) we obtain a unique cycle. The set of all cycles formed in this way, by adding separately each edge of G not contained in T, is the **fundamental set of cycles associated with T**. Sometimes we are not interested in the particular spanning forest chosen, and refer simply to a **fundamental set of cycles of G**. Note that the number of cycles in any fundamental set must equal the cycle rank of G. Figure 9.3 shows the fundamental set of cycles of the graph shown in Fig. 9.2 associated with the given spanning tree.

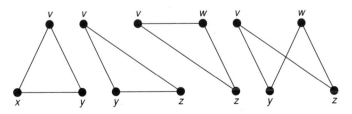

Fig. 9.3

In the light of our remarks at the end of Section 5, we may hope to be able to define a fundamental set of cutsets of a graph G associated with a spanning forest T. This is indeed the case. By Theorem 9.1(iv), the removal of any edge of T divides the vertex set of T into two disjoint sets V_1 and V_2. The set of all edges of G joining a vertex of V_1 to one of V_2 is a cutset of G, and the set of all cutsets obtained in this way, by removing separately each edge of T, is the **fundamental set of cutsets associated with T**. Note that the number of cutsets in any fundamental set must equal the cutset rank of G. The fundamental set of cutsets of the graph in Fig. 9.2 associated with the given spanning tree is $\{e_1, e_5\}$, $\{e_2, e_5, e_7, e_8\}$, $\{e_3, e_6, e_7, e_8\}$ and $\{e_4, e_6, e_8\}$.

Exercises 9

9.1ˢ   In the table of Fig. 2.9, locate all the trees.

9.2ˢ   Show, by drawing, that there are (up to isomorphism) six trees on 6 vertices and eleven trees on 7 vertices.

9.3   (i)   Prove that every tree is a bipartite graph.
    (ii)   Which trees are complete bipartite graphs?

9.4ˢ   Draw all the spanning trees in the graph of Fig. 9.4.

9.5   Draw all the spanning trees in the graph of Fig. 9.5.

Fig. 9.4

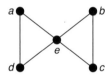

Fig. 9.5

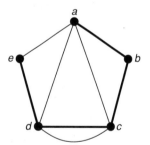

Fig. 9.6

9.6ˢ   Find the fundamental sets of cycles and cutsets of the graph in Fig. 9.6 associated with the spanning tree shown.

9.7   Find the cycle and cutset ranks of
    (i) K_5; (ii) $K_{3,3}$; (iii) W_5; (iv) N_5; (v) the Petersen graph.

9.8ˢ   Let G be a connected graph. What can you say about
    (i)   an edge of G that appears in every spanning tree?
    (ii)   an edge of G that appears in no spanning tree?

9.9   If G is a connected graph, a **centre** of G is a vertex v with the property that the maximum of the distances between v and the other vertices of G is as small as possible. By successively removing end-vertices, prove that every tree has either one centre or two adjacent centres. Give an example of a tree of each type with 7 vertices.

9.10* (i) Let C^* be a set of edges of a graph G. Show that, if C^* has an edge in common with each spanning forest of G, then C^* contains a cutset.

(ii) Obtain a corresponding result for cycles.

9.11 Let T_1 and T_2 be spanning trees of a connected graph G.

(i) If e is any edge of T_1, show that there exists an edge f of T_2 such that the graph $(T_1 - \{e\}) \cup \{f\}$ (obtained from T_1 on replacing e by f) is also a spanning tree.

(ii) Deduce that T_1 can be 'transformed' into T_2 by replacing the edges of T_1 one at a time by edges of T_2 in such a way that a spanning tree is obtained at each stage.

9.12* Show that if H and K are subgraphs of a graph G, and if $H \cup K$ and $H \cap K$ are defined in the obvious way, then the cutset rank ξ satisfies:

(i) $0 \le \xi(H) \le |E(H)|$ (the number of edges of H);

(ii) if H is a subgraph of K, then $\xi(H) \le \xi(K)$;

(iii) $\xi(H \cup K) + \xi(H \cap K) \le \xi(H) + \xi(K)$.

9.13* Let V be the vector space associated with a simple connected graph G, and let T be a spanning tree of G.

(i) Show that the fundamental set of cycles associated with T forms a basis for the cycle subspace W.

(ii) Obtain a corresponding result for the cutset subspace W^*.

(iii) Deduce that the dimensions of W and W^* are $\gamma(G)$ and $\xi(G)$, respectively.

10   Counting trees

The subject of graph enumeration is concerned with the problem of finding out how many non-isomorphic graphs possess a given property. The subject was initiated in the 1850s by Arthur Cayley, who later applied it to the problem of enumerating alkanes C_nH_{2n+2} with a given number of carbon atoms. As he realized, and as you will see in Section 11, this problem is that of counting the number of trees in which the degree of each vertex is either 4 or 1.

Many standard problems of graph enumeration have been solved. For example, it is possible to calculate the number of graphs, connected graphs, trees and Eulerian graphs with a given number of vertices and edges; corresponding general results for planar graphs and Hamiltonian graphs have, however, not yet been obtained. Most of the known results can be obtained by using a fundamental enumeration theorem due to Pólya, a good account of which may be found in Harary and Palmer [30]. Unfortunately, in almost every case it is impossible to express these results by means of simple formulas. For tables of some known results, see the Appendix.

This section is devoted primarily to two proofs of a famous result, usually attributed to Cayley, on the number of labelled trees with a given number of vertices. To see what is involved, consider Fig. 10.1, which shows three ways of labelling a tree with four vertices. Since the second labelled tree is the reverse of the first one, these two labelled trees are the same. On the other hand, neither is isomorphic to the third labelled tree, as you can see by comparing the degrees of vertex 3. Thus, the reverse of any labelling does not result in a new one, and so the number of ways of labelling this tree is $(4!)/2 = 12$. Similarly, the number of ways of labelling the tree in Fig. 10.2 is 4, since the central vertex can be labelled in four different ways, and each one determines the labelling. Thus, the total number of non-isomorphic labelled trees on four vertices is $12 + 4 = 16$.

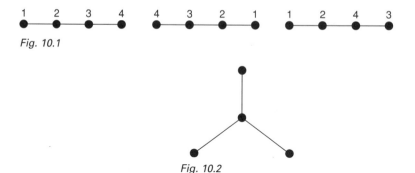

Fig. 10.1

Fig. 10.2

We now prove **Cayley's theorem**, which generalizes this result to labelled trees with n vertices.

THEOREM 10.1 (Cayley, 1889). *There are n^{n-2} distinct labelled trees on n vertices.*

Remark. The following proofs are due to Prüfer and Clarke; for other proofs, see Moon [31].

First proof. We establish a one-one correspondence between the set of labelled trees of order n and the set of sequences $(a_1, a_2, \ldots, a_{n-2})$, where each a_i is an integer satisfying $1 \le a_i \le n$. Since there are precisely n^{n-2} such sequences, the result follows immediately. We assume that $n \ge 3$, since the result is trivial if $n = 1$ or 2.

In order to establish the required correspondence, we first let T be a labelled tree of order n, and show how the sequence can be determined. If b_1 is the smallest label assigned to an end-vertex, we let a_1 be the label of the vertex adjacent to the vertex b_1. We then remove the vertex b_1 and its incident edge, leaving a labelled tree of order $n - 1$. We next let b_2 be the smallest label assigned to an end-vertex of our new tree, and let a_2 be the label of the vertex adjacent to the vertex b_2. We then remove the vertex b_2 and its incident edge, as before. We proceed in this way until there are only two vertices left, and the required sequence is $(a_1, a_2, \ldots, a_{n-2})$. For example, if T is the

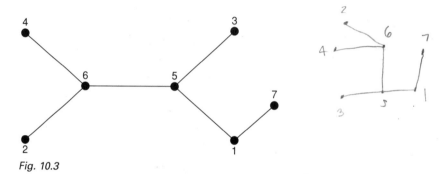

Fig. 10.3

labelled tree in Fig. 10.3, then $b_1 = 2$, $a_1 = 6$; $b_2 = 3$, $a_2 = 5$; $b_3 = 4$, $a_3 = 6$; $b_4 = 6$, $a_4 = 5$; $b_5 = 5$, $a_5 = 1$. The required sequence is therefore $(6, 5, 6, 5, 1)$.

To obtain the reverse correspondence, we take a sequence $(a_1, \ldots, a_{n-2})$, let b_1 be the smallest number that does *not* appear in it, and join the vertices a_1 and b_1. We then remove a_1 from the sequence, remove the number b_1 from consideration, and proceed as before. In this way we build up the tree, edge by edge. For example, if we start with the sequence $(6, 5, 6, 5, 1)$, then $b_1 = 2$, $b_2 = 3$, $b_3 = 4$, $b_4 = 6$, $b_5 = 5$, and the corresponding edges are 62, 53, 64, 56, 15. We conclude by joining the last two vertices not yet crossed out – in this case, 1 and 7. It is simple to check that if we start with any labelled tree, find the corresponding sequence, and then find the labelled tree corresponding to that sequence, then we obtain the tree we started from. We have therefore established the required correspondence, and the result follows. //

Second proof. Let $T(n, k)$ be the number of labelled trees on n vertices in which a given vertex v has degree k. We shall derive an expression for $T(n, k)$, and the result follows on summing from $k = 1$ to $k = n - 1$.

Let A be any labelled tree in which $\deg(v) = k - 1$. The removal from A of any edge wz that is not incident with v leaves two subtrees, one containing v and either w or z (w, say), and the other containing z. If we now join the vertices v and z, we obtain a labelled tree B in which $\deg(v) = k$ (see Fig. 10.4). We call a pair (A, B) of labelled trees a **linkage** if B can be obtained from A by the above construction. Our aim is to count the possible linkages (A, B).

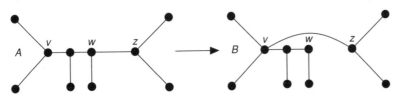

Fig. 10.4

Since A may be chosen in $T(n, k - 1)$ ways, and since B is uniquely defined by the edge wz which may be chosen in $(n - 1) - (k - 1) = n - k$ ways, the total number of linkages (A, B) is $(n - k)T(n, k - 1)$.

On the other hand, let B be a labelled tree in which $\deg(v) = k$, and let $T_1, \ldots, T_k$ be the subtrees obtained from B by removing the vertex v and each edge incident with v. Then we obtain a labelled tree A with $\deg(v) = k - 1$ by removing from B just one of these edges (vw_i, say, where w_i lies in T_i), and joining w_i to any vertex u of any other subtree T (see Fig. 10.5). Note that the corresponding pair (A, B) of labelled trees is a linkage, and that all linkages may be obtained in this way.

Since B can be chosen in $T(n, k)$ ways, and the number of ways of joining w_i to vertices in any other T_j is $(n - 1) - n_i$, where n_i is the number of vertices of T_i, the total number of linkages (A, B) is

$$T(n, k)\{(n - 1 - n_1) + \cdots + (n - 1 - n_k)\} = (n - 1)(k - 1)T(n, k),$$

since $n_1 + \cdots + n_k = n - 1$.

We have thus shown that

$$(n - k)T(n, k - 1) = (n - 1)(k - 1)T(n, k).$$

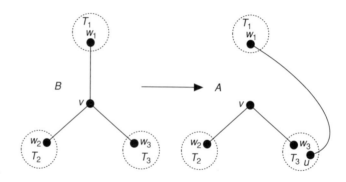

Fig. 10.5

On iterating this result, and using the obvious fact that $T(n, n - 1) = 1$, we deduce immediately that

$$T(n,k) = \binom{n-2}{k-1}(n-1)^{n-k-1} \qquad \binom{a}{b} = \frac{a!}{b!\,(a-b)!}$$

ex ponent

On summing over all possible values of k, we deduce that the number $T(n)$ of labelled trees on n vertices is given by

$$T(n) = \sum_{k=1}^{n-1} T(n,k) = \sum_{k=1}^{n-1} \binom{n-2}{k-1}(n-1)^{n-k-1}$$

$$= \{(n-1) + 1\}^{n-2} = n^{n-2}. \; //$$

COROLLARY 10.2. *The number of spanning trees of K_n is n^{n-2}.*

Proof. To each labelled tree on n vertices there corresponds a unique spanning tree of K_n. Conversely, each spanning tree of K_n gives rise to a unique labelled tree on n vertices. //

We conclude this section by stating an important result that can be used to calculate the number of spanning trees in any connected simple graph. It is called the **matrix-tree theorem** and its proof may be found in Harary [9].

THEOREM 10.3. *Let G be a connected simple graph with vertex set $\{v_1, \ldots, v_n\}$, and let $\mathbf{M} = (m_{ij})$ be the $n \times n$ matrix in which $m_{ii} = \deg(v_i)$, $m_{ij} = -1$ if v_i and v_j are adjacent, and $m_{ij} = 0$ otherwise. Then the number of spanning trees of G is equal to the cofactor of any element of $\mathbf{M}$.*

Exercises 10

10.1ˢ   Verify directly that there are exactly 125 labelled trees on 5 vertices.

10.2ˢ   In the first proof of Cayley's theorem, find:

(i)   the labelled trees corresponding to the sequences $(1, 2, 3, 4)$ and $(3, 3, 3, 3)$;

(ii) the sequences corresponding to the labelled trees in Fig. 10.6.

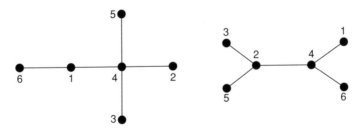

Fig. 10.6

10.3   (i)   Find the number of trees on n vertices in which a given vertex is an end-vertex.

(ii) Deduce that, if n is large, then the probability that a given vertex of a tree with n vertices is an end-vertex is approximately e^{-1}.

10.4ˢ   How many spanning trees has $K_{2,s}$?

10.5   Let $\tau(G)$ be the number of spanning trees in a connected graph G.

(i)   Prove that, for any edge e, $\tau(G) = \tau(G - e) + \tau(G\backslash e)$.

(ii) Use this result to calculate $\tau(K_{2,3})$.

10.6*   Use the matrix-tree theorem to prove Cayley's theorem.

10.7*   Let $T(n)$ be the number of labelled trees on n vertices.

(i)   By counting the number of ways of joining a labelled tree on k vertices and one on $n - k$ vertices, prove that

$$2(n-1)T(n) = \sum_{k=1}^{n-1} \binom{n}{k} k(n-k)T(k)T(n-k).$$

(ii) Deduce the identity

$$\sum_{k=1}^{n-1} \binom{n}{k} k^{k-1}(n-k)^{n-k-1} = 2(n-1)n^{n-2}.$$

11   *More applications*

In Section 8 we considered three problems that arise in operational research – the shortest path problem, the Chinese postman problem and the travelling salesman problem. In this section we consider four further graph theory applications, taken from operational research, organic chemistry, electrical network theory and computing, and each involving the use of trees.

The minimum connector problem

Let us suppose that we wish to build a railway network connecting n given cities so that a passenger can travel from any city to any other. If, for economic reasons, the total amount of track must be a minimum, then the graph formed by taking the n cities as vertices and the connecting rails as edges must be a tree. The problem is to find an efficient algorithm for deciding which of the n^{n-2} possible trees connecting these cities uses the least amount of track, assuming that the distances between all the pairs of cities are known.

As before, we can reformulate the problem in terms of weighted graphs. We denote the weight of the edge e by $w(e)$, and our aim is to find the spanning tree T with least possible total weight $W(T)$. Unlike some of the problems we considered earlier, there is a simple algorithm that provides the solution. It is known as a **greedy algorithm**, and involves choosing edges of minimum weight in such a way that no cycle is created. For example, if there are five cities, as shown in Fig. 11.1, then we start by choosing the edges AB (weight 2) and BD (weight 3). We cannot then choose the edge AD (weight 4), since it would create the cycle ABD, so we choose the edge DE (weight 5). We cannot then choose the edges AE or BE (weight 6), since each would create a cycle, so we choose the edge BC (weight 7). This completes the tree (see Fig. 11.2).

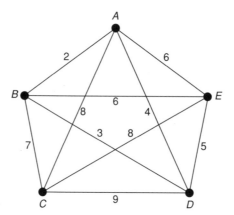

Fig. 11.1

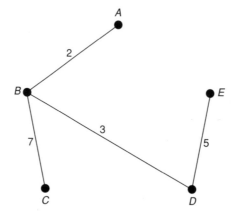

Fig. 11.2

The algorithm is described in general in the following theorem.

> **THEOREM 11.1.** *Let G be a connected graph with n vertices. Then the following construction gives a solution of the minimum connector problem:*
> (i) *let e_1 be an edge of G of smallest weight;*
> (ii) *define $e_2, e_3, \ldots, e_{n-1}$ by choosing at each stage a new edge of smallest possible weight that forms no cycle with the previous edges e_i. The required spanning tree is the subgraph T of G whose edges are $e_1, \ldots, e_{n-1}$.*

Proof. The fact that T is a spanning tree of G follows immediately from Theorem 9.1(ii). It remains only to show that the total weight of T is a minimum. In order to do this, suppose that S is a spanning tree of G such that $W(S) < W(T)$. If e_k is the first edge in the above sequence that does not lie in S, then the subgraph of T formed by adding e_k to S contains a unique cycle C containing the edge e_k. Since C contains an edge e lying in S but not in T, the subgraph obtained from S on replacing e by e_k is a spanning tree S'. But by the construction, $w(e_k) \le w(e)$, and so $W(S') \le W(S)$, and S' has one more edge in common with T than S. It follows on repeating this procedure that we can change S into T, one step at a time, with the total weight decreasing at each stage. Hence $W(T) \le W(S)$, giving the required contradiction. //

We now apply the greedy algorithm to obtain a lower bound for the solution of the travelling salesman problem. This is useful, since the greedy algorithm is an efficient algorithm, whereas no efficient general algorithms are known for the travelling salesman problem.

If we take any Hamiltonian cycle in a weighted complete graph and remove any vertex v, then we obtain a semi-Hamiltonian path, and such a path must be a spanning tree. So any solution of the travelling salesman problem must consist of a spanning tree of this type together with two edges incident to v. It follows that if we take the weight of a *minimum-weight* spanning tree (obtained by the greedy algorithm) and add the two *smallest* weights of edges incident with v, then we get a *lower bound* for the solution of the travelling salesman problem.

For example, if we take the weighted graph of Fig. 11.1 and remove the vertex C, then the remaining weighted graph has the four vertices A, B, D and E. The minimum-weight spanning tree joining these four vertices is the tree whose edges are AB, BD and DE, with total weight 10, and the two edges of minimum weight incident with C are CB and CA (or CE) with total weight 15 (see Fig. 11.3). Thus, the required lower bound for the travelling salesman problem is 25. Since the correct answer is 26, we

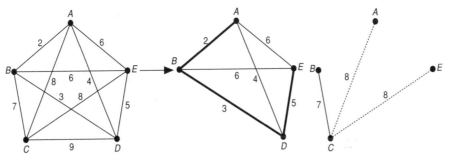

Fig. 11.3

see that this approach to the travelling salesman problem can yield surprisingly good results.

Enumeration of chemical molecules

One of the earliest uses of trees was in the enumeration of chemical molecules. If we have a molecule consisting only of carbon atoms and hydrogen atoms, then we can represent it as a graph in which each carbon atom appears as a vertex of degree 4, and each hydrogen atom appears as a vertex of degree 1.

The graphs of *n*-butane and 2-methyl propane are shown in Fig. 11.4. Although they have the same chemical formula C_4H_{10}, they are different molecules because the atoms are arranged differently within the molecule. These two molecules form part of a general class of molecules known as the *alkanes*, or *paraffins*, with chemical formula C_nH_{2n+2}, and it is natural to ask how many different molecules there are with this formula.

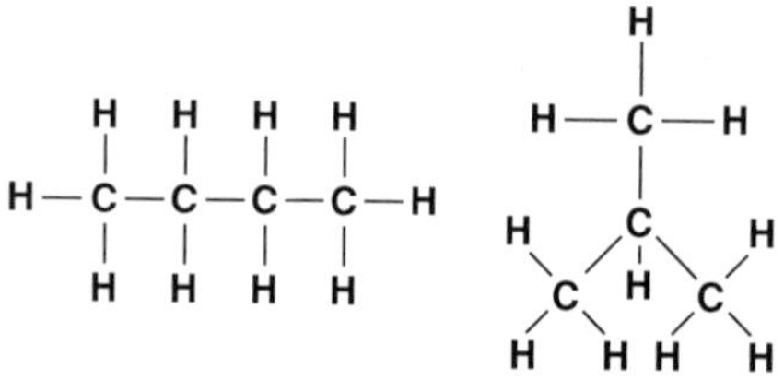

Fig. 11.4

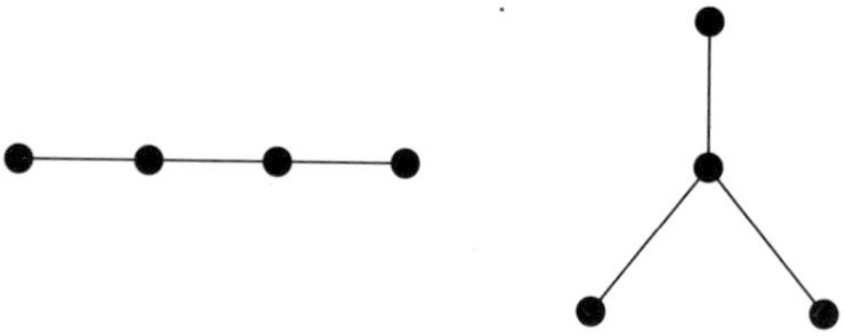

Fig. 11.5

To answer this, we note first that the graph of any molecule with formula C_nH_{2n+2} is a tree, by Theorem 9.1(iii), since it is connected and has $n + (2n + 2) = 3n + 2$ vertices and $\{4n + (2n + 2)\}/2 = 3n + 1$ edges. Note also that the molecule is determined completely once we know how the carbon atoms are arranged, since hydrogen atoms can then be added in such a way as to bring the degree of each carbon vertex to 4. We can thus discard the hydrogen atoms, as in Fig. 11.5, and the problem reduces to that of finding the number of trees with *n* vertices, each of degree 4 or less.

This problem was solved by Cayley in 1875, by counting the number of ways in which trees can be built up from their centre(s) (see Exercise 9.9). The details of this argument are too complicated to describe here, but may be found in Biggs, Lloyd and Wilson [11]. Much of Cayley's work has been superseded by G. Pólya and others, with the result that many chemical series have been enumerated by graph-theoretical techniques.

Electrical networks

Suppose that we are given the electrical network in Fig. 11.6, and that we wish to find the current in each wire.

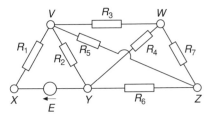

Fig. 11.6

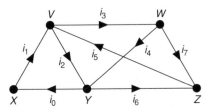

Fig. 11.7

To do this, we assign an arbitrary direction to the current in each wire, as in Fig. 11.7, and apply *Kirchhoff's laws*:

(i) the algebraic sum of the currents at each vertex is 0;
(ii) the total voltage in each cycle is obtained by adding the products of the currents i_k and resistances R_k in that cycle.

Applying Kirchhoff's second law to the cycles *VYXV*, *VWYV* and *VWYXV*, we obtain the equations

$$i_1R_1 + i_2R_2 = E; \quad i_3R_3 + i_4R_4 - i_2R_2 = 0; \quad i_1R_1 + i_3 R_3 + i_4R_4 = E.$$

The last of these three equations is simply the sum of the first two, and gives us no further information. Similarly, if we have the Kirchhoff equations for the cycles *VWYV* and *WZYW*, then we can deduce the equation for the cycle *VWZYV*. It will save a lot of work if we can find a set of cycles that gives us the information we need without any redundancy, and this can be done by using a fundamental set of cycles (see Section 9). In this example, taking the fundamental system of cycles in Fig. 9.3, we obtain the following equations:

for the cycle *VYXV*,	$i_1R_1 + i_2R_2$	$= E,$
for the cycle *VYZV*,	$i_2R_2 + i_5R_5 + i_6R_6$	$= 0,$
for the cycle *VWZV*,	$i_3R_3 + i_5R_5 + i_7R_7$	$= 0,$
for the cycle *VYWZV*,	$i_2R_2 - i_4R_4 + i_5R_5 + i_7R_7$	$= 0.$

The equations arising from Kirchhoff's first law are:

for the vertex *X*,	$i_0 - i_1$	$= 0,$
for the vertex *V*,	$i_1 - i_2 - i_3 + i_5$	$= 0,$
for the vertex *W*,	$i_3 - i_4 - i_7$	$= 0,$
for the vertex *Z*,	$i_5 - i_6 - i_7$	$= 0.$

These eight equations can now be solved to give the eight currents $i_0, \ldots, i_7$. For example, if $E = 12$, and if each wire has unit resistance (that is, $R_i = 1$ for each i), then the solution is as given in Fig. 11.8.

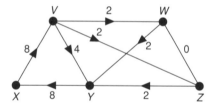

Fig. 11.8

Searching trees

In many applications, the trees that we consider have a hierarchical structure, with one vertex at the top (called the **root**), and the other vertices branching down from it, as in Fig. 11.9. For example, a computer file or a library classification system is often organized in this way, with information stored at the vertices.

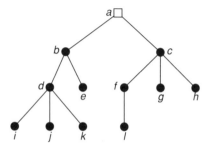

Fig. 11.9

If a particular piece of information is required, we need to be able to search the tree in a systematic way. This often involves examining every part of the tree until the desired vertex is found. We should like to find a search technique that eventually visits all parts of the tree without visiting any vertex too often.

There are two well-known search procedures – **depth first search** and **breadth first search**. Both methods visit all the vertices, but in a different order. No rule can be given for which method should be used for a particular problem – each has its advantages. For example, a breadth first search is used in the shortest path algorithm (Section 8), whereas a depth first search is used for finding network flows (Section 29).

In breadth first search, we fan out to as many vertices as possible, before penetrating deeper into the tree. This means that we visit all the vertices adjacent to the current vertex before proceeding to another vertex. For example, consider the tree in Fig. 11.9. In order to perform a breadth first search, we start at vertex a and visit the vertices b and c that are adjacent to a. We then visit the vertices d and e adjacent to b, and the vertices f, g and h adjacent to c. Finally we visit the vertices i, j and k adjacent to d, and l adjacent to f. This gives us the labelled tree in Fig. 11.10, where the labels correspond to the order in which the vertices are visited.

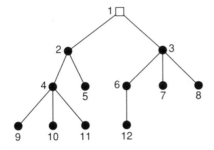

Fig. 11.10

In depth first search, we penetrate as deeply as possible into a tree before fanning out to other vertices. For example, consider again the tree in Fig. 11.9. In order to perform a depth first search, we start at vertex *a* and move down to *b*, *d* and *i*. Since we cannot penetrate further, we backtrack to *d* and then go down to *j*. We must then backtrack again, and go to *k*. We now backtrack via *d* to *b*, from which we can go down to *e*. Backtracking to *a* then takes us to *c*, *f* and *l*, and eventually to *g* and *h* before returning to *a*. This gives us the labelled tree in Fig. 11.11, where the labels correspond to the order in which the vertices are visited.

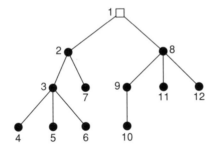

Fig. 11.11

Exercises 11

11.1ˢ   Use the greedy algorithm to find a minimum-weight spanning tree in the graph in Fig. 11.12.

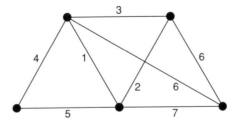

Fig. 11.12

11.2   Find a minimum-weight spanning tree in the graph in Fig. 11.13.

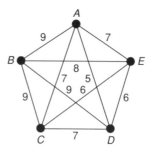

Fig. 11.13

11.3   Show that if each edge of a connected weighted graph G has the same weight, then the greedy algorithm gives a method for constructing a spanning tree in G.

11.4   (i)   How would you adapt the greedy algorithm to find a *maximum*-weight spanning tree?
         (ii) Find a maximum-weight spanning tree for each of the weighted graphs in Figs. 11.1 and 11.12.

11.5ˢ   In applying the greedy algorithm to the travelling salesman problem in the text, what lower bounds do you get if you remove each of the vertices A, B, D and E, instead of C?

11.6ˢ   Show that, for each value of n, the graph associated with the alcohol $C_nH_{2n+1}OH$ is a tree (the oxygen vertex has degree 2).

11.7   Draw the chemical molecules with the formulas C_5H_{12} and C_6H_{14}.

11.8ˢ   Perform a breadth first search and a depth first search on the tree in Fig. 11.14.

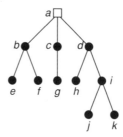

Fig. 11.14

11.9   Perform a breadth first search and a depth first search on the tree in Fig. 11.15.

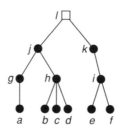

Fig. 11.15

11.10[s] Verify the currents in Fig. 11.8, by applying Kirchhoff's laws to the fundamental cycles associated with the spanning tree with edges VX, VW, WZ and YZ.

11.11* Write down and solve Kirchhoff's equations for the network of Fig. 11.16, in which the numbers are resistances.

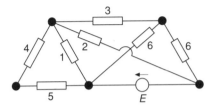

Fig. 11.16

Chapter 5

Planarity

Flattery will get you nowhere.
Popular saying

We now embark upon a study of topological graph theory, in which graphs become tied up with topological notions such as planarity, genus, etc. In particular, we investigate when a graph can be drawn in the plane and on other surfaces. In Section 12, we discuss planar graphs, prove that some graphs are not planar, and state Kuratowski's characterization of planar graphs. In Section 13, we prove Euler's formula relating the numbers of vertices, edges and faces of a graph drawn in the plane, and generalize it to graphs drawn on other surfaces in Section 14. In Section 15 we study duality, and the chapter concludes, in Section 16, with some material on infinite graphs.

12   Planar graphs

A **planar graph** is a graph that can be drawn in the plane without crossings – that is, so that no two edges intersect geometrically except at a vertex to which both are incident. Any such drawing is a **plane drawing**. For convenience, we often use the abbreviation **plane graph** for a plane drawing of a planar graph. For example, Fig. 12.1 shows three drawings of the planar graph K_4, but only the second and third are plane graphs.

Fig. 12.1

The right-hand drawing in Fig. 12.1 leads us to ask whether every planar graph can be drawn in the plane so that each edge is represented by a straight line. Although loops or multiple edges cannot be drawn as straight lines, it was proved independently by K. Wagner in 1936 and I. Fáry in 1948 that *every simple planar graph can be drawn with straight lines*; see Chartrand and Lesniak [8] for details.

Not all graphs are planar, as the following theorem shows:

THEOREM 12.1. $K_{3,3}$ *and* K_5 *are non-planar.*

Remark. We give two proofs of this result. The first one, presented here, is constructive. The second proof, which we give in Section 13, appears as a corollary of Euler's formula.

Proof. Suppose first that $K_{3,3}$ is planar. Since $K_{3,3}$ has a cycle $u \to v \to w \to x \to y \to z \to u$ of length 6, any plane drawing must contain this cycle drawn in the form of a hexagon, as in Fig. 12.2.

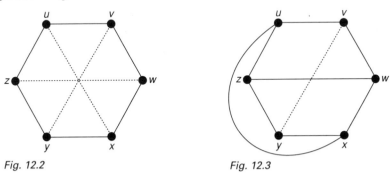

Fig. 12.2   Fig. 12.3

Now the edge wz must lie either wholly inside the hexagon or wholly outside it. We deal with the case in which wz lies inside the hexagon – the other case is similar. Since the edge ux must not cross the edge wz, it must lie outside the hexagon; the situation is now as in Fig. 12.3. It is then impossible to draw the edge vy, as it would cross either ux or wz. This gives the required contradiction.

Now suppose that K_5 is planar. Since K_5 has a cycle $v \to w \to x \to y \to z \to v$ of length 5, any plane drawing must contain this cycle drawn in the form of a pentagon, as in Fig. 12.4.

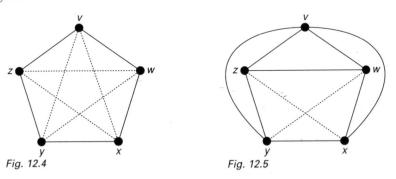

Fig. 12.4   Fig. 12.5

Now the edge wz must lie either wholly inside the pentagon or wholly outside it. We deal with the case in which wz lies inside the pentagon – the other case is similar. Since the edges vx and vy do not cross the edge wz, they must both lie outside the pentagon; the situation is now as in Fig. 12.5. But the edge xz cannot cross the edge vy and so must lie inside the pentagon; similarly the edge wy must lie inside the pentagon, and the edges wy and xz must then cross. This gives the required contradiction. //

Note that every subgraph of a planar graph is planar, and that every graph with a non-planar subgraph must be non-planar. It follows that any graph with $K_{3,3}$ or K_5 as a subgraph is non-planar. In fact, as we shall see, these two graphs are the 'building blocks' for non-planar graphs, in the sense that every non-planar graph must 'contain' at least one of them.

To make this statement more precise, we define two graphs to be **homeomorphic** if both can be obtained from the same graph by inserting new vertices of degree 2 into its edges. For example, any two cycle graphs are homeomorphic, as are the graphs of Fig. 12.6.

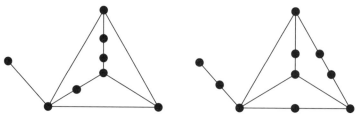

Fig. 12.6

Note that the introduction of the term 'homeomorphic' is merely a technicality, as the insertion or deletion of vertices of degree 2 is irrelevant to considerations of planarity. However, it enables us to state the following important result, known as **Kuratowski's theorem**, which gives a necessary and sufficient condition for a graph to be planar.

THEOREM 12.2 (Kuratowski, 1930). *A graph is planar if and only if it contains no subgraph homeomorphic to K_5 or $K_{3,3}$.*

The proof of Kuratowski's theorem is long and involved, and we omit it; see Bondy and Murty [7] or Harary [9] for a proof. We shall, however, use Kuratowski's theorem to obtain another criterion for planarity. To do so, we first define a graph H to be **contractible** to K_5 or $K_{3,3}$ if we can obtain K_5 or $K_{3,3}$ by successively contracting edges of H. For example, the Petersen graph is contractible to K_5, as we can see by contracting the five 'spokes' joining the inner and outer 5-cycles (see Fig. 12.7).

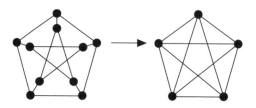

Fig. 12.7

THEOREM 12.3. *A graph is planar if and only if it contains no subgraph contractible to K_5 or $K_{3,3}$.*

Sketch of proof. ⇐ Assume first that the graph G is non-planar. Then, by Kuratowski's theorem, G contains a subgraph H homeomorphic to K_5 or $K_{3,3}$. On successively contracting edges of H that are incident to a vertex of degree 2, we see that H is contractible to K_5 or $K_{3,3}$.

⇒ Now assume that G contains a subgraph H contractible to $K_{3,3}$, and let the vertex v of $K_{3,3}$ arise from contracting the subgraph H_v of H (see Fig. 12.8).

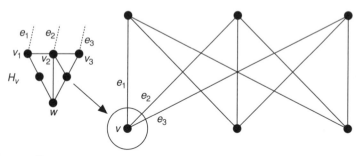

Fig. 12.8

The vertex v is incident in $K_{3,3}$ to three edges e_1, e_2 and e_3. When regarded as edges of H, these edges are incident to three (not necessarily distinct) vertices v_1, v_2 and v_3 of H_v. If v_1, v_2 and v_3 are distinct, then we can find a vertex w of H_v and three paths from w to these vertices, intersecting only at w. (There is a similar construction if the vertices are not distinct, the paths degenerating in this case to single vertices.) It follows that we can replace the subgraph H_v by a vertex w and three paths leading out of it. If this construction is carried out for each vertex of $K_{3,3}$, and the resulting paths joined up with the corresponding edges of $K_{3,3}$, then the resulting subgraph is homeomorphic to $K_{3,3}$. It follows from Kuratowski's theorem that G is non-planar (see Fig. 12.9).

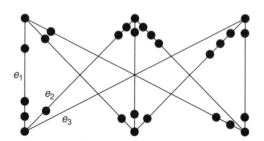

Fig. 12.9

A similar argument can be carried out if G contains a subgraph contractible to K_5. Here the details are more complicated, as the subgraph we obtain by the above process can be homeomorphic to either K_5 or $K_{3,3}$; see Chartrand and Lesniak [8]. //

We conclude this section by introducing the 'crossing-number' of a graph. If we try to draw K_5 or $K_{3,3}$ on the plane, then there must be at least one crossing of edges, since these graphs are not planar. However, we do not need more than one crossing (see Fig. 12.10), and we say that K_5 and $K_{3,3}$ have crossing number 1.

More generally, the **crossing number** $\mathrm{cr}(G)$ of a graph G is the minimum number of crossings that can occur when G is drawn in the plane. Thus, the crossing number

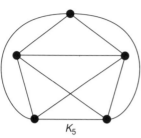
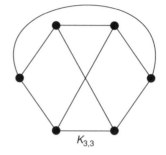

K_5 $K_{3,3}$

Fig. 12.10

can be used to measure how 'unplanar' G is. For example, the crossing number of any planar graph is 0, and $\text{cr}(K_5) = \text{cr}(K_{3,3}) = 1$. Note that the word 'crossing' always refers to the intersection of just two edges; crossings of three or more edges are not permitted.

Exercises 12

12.1[s]   Show, by drawing, that the following graphs are planar:

    (i)   the wheel W_5;

    (ii)   the graph of the octahedron.

12.2   Show how the graph of Fig. 12.11 can be drawn in the plane without crossings.

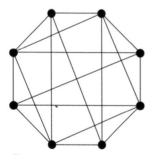

Fig. 12.11

12.3[s]   Three unfriendly neighbours use the same water, oil and treacle wells. In order to avoid meeting, they wish to build non-crossing paths from each of their houses to each of the three wells. Can this be done?

12.4[s]   Which complete graphs and complete bipartite graphs are planar?

12.5   (i)   For which values of k is the k-cube Q_k planar?

    (ii)   For which values of r, s and t is the complete tripartite graph $K_{r,s,t}$ planar?

12.6   Prove that the Petersen graph is non-planar

    (i)   by using Theorem 12.2;

    (ii)   by using Theorem 12.3.

    (Hint for part (i): remove the two 'horizontal' edges.)

12.7[s]   Give an example of

    (i)   a non-planar graph that is not homeomorphic to K_5 or $K_{3,3}$;

    (ii)   a non-planar graph that is not contractible to K_5 or $K_{3,3}$.

    Why does the existence of these graphs not contradict Theorems 12.2 and 12.3?

12.8   Given that two homeomorphic graphs have n_i vertices and m_i edges ($i = 1, 2$), show that $m_1 - n_1 = m_2 - n_2$.

12.9   A graph G is **outerplanar** if G can be drawn in the plane so that all of its vertices lie on the exterior boundary.
   (i)   Show that K_4 and $K_{2,3}$ are not outerplanar.
   (ii)   Deduce that, if G is an outerplanar graph, then G contains no subgraph homeomorphic or contractible to K_4 or $K_{2,3}$. (The converse result also holds, yielding a Kuratowski-type criterion for a graph to be outerplanar.)

12.10^s   Show that $K_{4,3}$ and the Petersen graph each has crossing number 2.

12.11*   Given that r and s are both even, prove that

$$\mathrm{cr}(K_{r,s}) \le rs(r-2)(s-2)/16,$$

and obtain corresponding results when r and/or s is odd.
(Hint: place the r vertices along the x-axis with $r/2$ vertices on each side of the origin, place the s vertices along the y-axis in a similar way, and count the crossings.)

12.12*   Let G be a planar graph with vertex set $\{v_1, \ldots, v_n\}$, and let $p_1, \ldots, p_n$ be any n distinct points in the plane. Give a heuristic argument to show that G can be drawn in the plane in such a way that the point p_i represents the vertex v_i, for each i.

12.13*   By placing the vertices at the points $(1, 1^2, 1^3), (2, 2^2, 2^3), (3, 3^2, 3^3), \ldots$, prove that any simple graph can be drawn without crossings in Euclidean three-dimensional space so that each edge is represented by a straight line.

13   Euler's formula

If G is a planar graph, then any plane drawing of G divides the set of points of the plane not lying on G into regions, called **faces**. For example, the plane graphs in Figs 13.1 and 13.2 have eight faces and four faces, respectively. Note that, in each case, the face f_4 is unbounded; it is called the **infinite face**.

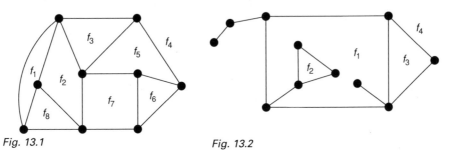

Fig. 13.1   Fig. 13.2

There is nothing special about the infinite face – in fact, any face can be chosen as the infinite face. To see this, we map the graph onto the surface of a sphere by stereographic projection (see Fig. 13.3). We now rotate the sphere so that the point of projection (the north pole) lies inside the face we want as the infinite face, and then project the graph down onto the plane tangent to the sphere at the south pole. The chosen face is now the infinite face.

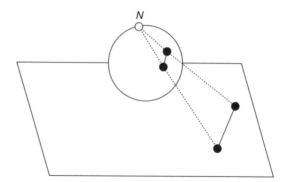

Fig. 13.3

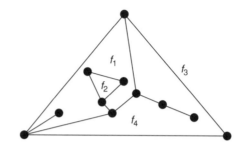

Fig. 13.4

Figure 13.4 shows a representation of the graph of Fig. 13.2 in which the infinite face is f_3.

We now state and prove **Euler's formula** that tells us that whatever plane drawing of a planar graph we take, the number of faces is always the same and is given by a simple formula. An alternative proof is outlined in Exercise 13.11.

> **THEOREM 13.1** (Euler, 1750). *Let G be a plane drawing of a connected planar graph, and let n, m and f denote respectively the number of vertices, edges and faces of G. Then*
> $$n - m + f = 2.$$

Remark. An example of this theorem is given by Fig. 13.2, where $n = 11$, $m = 13$, $f = 4$, and $n - m + f = 11 - 13 + 4 = 2$.

Proof. The proof is by induction on the number of edges of G. If $m = 0$, then $n = 1$ (since G is connected) and $f = 1$ (the infinite face). The theorem is therefore true in this case.

Now suppose that the theorem holds for all graphs with at most $m - 1$ edges, and let G be a graph with m edges. If G is a tree, then $m = n - 1$ and $f = 1$, so that $n - m + f = 2$, as required. If G is not a tree, let e be an edge in some cycle of G. Then $G - e$ is a connected plane graph with n vertices, $m - 1$ edges, and $f - 1$ faces, so that $n - (m - 1) + (f - 1) = 2$, by the induction hypothesis. It follows that $n - m + f = 2$, as required. //

This result is often called 'Euler's polyhedron formula', since it relates the numbers of vertices, edges and faces of a convex polyhedron. For example, for a cube we have $n = 8$, $m = 12$, $f = 6$, and $n - m + f = 8 - 12 + 6 = 2$ (see Fig. 13.5).

Fig. 13.5

To see the connection in general, project the polyhedron out onto its circumsphere, and then use stereographic projection (as in Fig. 13.3) to project it down onto the plane. The resulting graph is a 3-connected plane graph in which each face is bounded by a polygon – such a graph is called a **polyhedral graph** (see Fig. 13.1). For convenience, we restate Theorem 13.1 for such graphs.

COROLLARY 13.2. *Let G be a polyhedral graph. Then, with the above notation,*
$n - m + f = 2$.

Euler's formula can easily be extended to disconnected graphs:

COROLLARY 13.3. *Let G be a plane graph with n vertices, m edges, f faces and k components. Then*
$n - m + f = k + 1$.

Proof. The result follows on applying Euler's formula to each component separately, remembering not to count the infinite face more than once. //

All the results mentioned so far in this section apply to arbitrary plane graphs. We now restrict ourselves to simple graphs.

COROLLARY 13.4. *(i) If G is a connected simple planar graph with n (≥ 3) vertices and m edges, then $m \leq 3n - 6$.*
(ii) If, in addition, G has no triangles, then $m \leq 2n - 4$.

Proof. (i) We can assume that we have a plane drawing of G. Since each face is bounded by at least three edges, it follows on counting up the edges around each face that $3f \leq 2m$; the factor 2 appears since each edge bounds two faces. We obtain the required result by combining this inequality with Euler's formula.
(ii) This part follows in a similar way, except that the inequality $3f \leq 2m$ is replaced by $4f \leq 2m$. //

Using this corollary, we can give an alternative proof of Theorem 12.1.

COROLLARY 13.5. K_5 *and* $K_{3,3}$ *are non-planar.*

Proof. If K_5 is planar then, applying Corollary 13.4(i), we obtain $10 < 9$, which is a contradiction. If $K_{3,3}$ is planar then, applying Corollary 13.4(ii), we obtain $9 < 8$, which is also a contradiction. //

We use a similar argument to prove the following theorem, which will be useful when we study the colouring of graphs.

THEOREM 13.6. *Every simple planar graph contains a vertex of degree at most 5.*

Proof. Without loss of generality we can assume the graph to be connected, and to have at least three vertices. If each vertex has degree at least 6, then, with the above notation, we have $6n \leq 2m$, and so $3n \leq m$. It follows immediately from Corollary 13.4(i) that $3n \leq 3n - 6$, which is a contradiction. //

We conclude this section with a few remarks on the 'thickness' of a graph. In electrical engineering, parts of networks are sometimes printed on one side of a non-conducting plate, and are called 'printed circuits'. Since the wires are not insulated, they must not cross and the corresponding graphs must be planar (see Fig. 13.6).

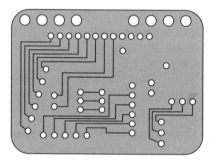

Fig. 13.6

For a general network, we may need to know how many printed circuits are needed to complete the entire network. To this end, we define the **thickness** $t(G)$ of a graph G to be the smallest number of planar graphs that can be superimposed to form G. Like the crossing number, the thickness is a measure of how 'unplanar' a graph is; for example, the thickness of a planar graph is 1, and of K_5 and $K_{3,3}$ is 2. Figure 13.7 shows that the thickness of K_6 is 2.

As we shall see, a lower bound for the thickness of a graph is easily obtained from Euler's formula. Surprisingly, this trivial lower bound frequently turns out to be the

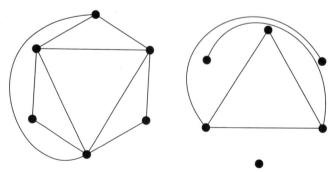

Fig. 13.7

correct value, as we can verify by direct construction. In deriving this lower bound, we use the symbols $\lfloor x \rfloor$ and $\lceil x \rceil$ to denote respectively the largest integer not greater than x and the smallest integer not less than x; for example, $\lfloor 3 \rfloor = \lceil 3 \rceil = 3$; $\lfloor \pi \rfloor = 3$; $\lceil \pi \rceil = 4$.

> **THEOREM 13.7.** *Let G be a simple graph with n (≥ 3) vertices and m edges. Then the thickness t(G) of G satisfies the inequalities*
>
> $$t(G) \geq \lceil m/(3n-6) \rceil \text{ and } t(G) \geq \lfloor (m+3n-7)/(3n-6) \rfloor.$$

Proof. The first part is an immediate application of Corollary 13.4(i), the brackets arising from the fact that the thickness must be an integer. The second part follows from the first by using the easily proved relation $\lceil a/b \rceil = \lfloor (a+b-1)/b \rfloor$, where a and b are positive integers. //

Exercises 13

13.1[s] Verify Euler's formula for
  (i)   the wheel W_8;
  (ii) the graph of the octahedron;
  (iii) the graph of Fig. 13.1;
  (iv) the complete bipartite graph $K_{2,7}$.

13.2 Redraw the graph of Fig. 13.2 with
  (i) f_1 as the infinite face;
  (ii) f_2 as the infinite face.

13.3[s] (i)   Use Euler's formula to prove that, if G is a connected planar graph of girth 5 then, with the above notation, $m \leq 5(n-2)/3$. Deduce that the Petersen graph is non-planar.
  (ii) Obtain an inequality, generalizing that in part (i), for connected planar graphs of girth r.

13.4 Let G be a polyhedron (or polyhedral graph), each of whose faces is bounded by a pentagon or a hexagon.
  (i)   Use Euler's formula to show that G must have at least 12 pentagonal faces.
  (ii) Prove, in addition, that if there are exactly three faces meeting at each vertex, then G has exactly 12 pentagonal faces.

13.5 Let G be a simple plane graph with fewer than 12 faces, in which each vertex has degree at least 3.
  (i)   Use Euler's formula to prove that G has a face bounded by at most four edges.
  (ii) Give an example to show that the result of part (i) is false if G has 12 faces.

13.6 (i)   Let G be a simple connected cubic plane graph, and let φ_k be the number of k-sided faces. By counting the number of vertices and edges of G, prove that
    $$3\varphi_3 + 2\varphi_4 + \varphi_5 - \varphi_7 - 2\varphi_8 - \cdots = 12.$$
  (ii) Deduce that G has at least one face bounded by at most five edges.

13.7 Let G be a simple graph with at least 11 vertices, and let $\bar{G}$ be its complement.
  (i)   Prove that G and $\bar{G}$ cannot both be planar.
    (In fact, a similar result holds if 11 is replaced by 9.)
  (ii) Find a graph G with 8 vertices such that G and $\bar{G}$ are both planar.

13.8[s]   Find the thickness of
   (i)   the Petersen graph;
   (ii)   the 4-cube Q_4.

13.9   (i)   Show that the thickness of K_n satisfies $t(K_n) \geq \lfloor (n+7)/6 \rfloor$.
   (ii)   Use the results of Exercise 13.7 to prove that equality holds if $n = 8$, but not if $n = 9$ or 10.
   (In fact, equality holds for all n other than 9 or 10.)

13.10   (i)   Use Corollary 13.4(ii) to prove that
   $$t(K_{r,s}) \geq \lceil rs/(2r + 2s - 4) \rceil,$$
   and verify that equality holds for $t(K_{3,3})$.
   (ii)   Given that r is even, show that $t(K_{r,s}) = r$, and deduce from part (i) that $t(K_{r,s}) = r/2$ if $s > (r-2)^2/2$.

13.11*   Let G be a polyhedral graph and let W be the cycle subspace of G.
   (i)   Show that the polygons bounding the finite faces of G form a basis for W.
   (ii)   Deduce Corollary 13.2.

14   Graphs on other surfaces

In the previous two sections we considered graphs drawn in the plane or (equivalently) on the surface of a sphere. We now consider drawing graphs on other surfaces – for example, the torus. It is easy to see that K_5 and $K_{3,3}$ can be drawn without crossings on the surface of a torus (see Fig. 14.1), and it is natural to ask whether there are analogues of Euler's formula and Kuratowski's theorem for graphs drawn on such surfaces.

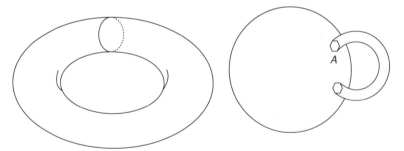

Fig. 14.1

The torus can be thought of as a sphere with one 'handle'. More generally, a surface is of **genus** g if it is topologically homeomorphic to a sphere with g handles. If you are unfamiliar with these terms, just think of graphs drawn on the surface of a dough-nut with g holes in it. Thus the genus of a sphere is 0, and that of a torus is 1.

A graph that can be drawn without crossings on a surface of genus g, but not on one of genus $g - 1$, is a **graph of genus** g. Thus, K_5 and $K_{3,3}$ are graphs of genus 1 (also called **toroidal graphs**).

The following result gives us an upper bound for the genus of a graph.

THEOREM 14.1. *The genus of a graph does not exceed the crossing number.*

Proof. We draw the graph on the surface of a sphere so that the number of crossings is as small as possible, and is therefore equal to the crossing number c. At each crossing, we construct a 'bridge' (as in Fig. 1.1) and run one edge over the bridge and the other under it. Since each bridge can be thought of as a handle, we have drawn the graph on the surface of a sphere with c handles. It follows that the genus does not exceed c. //

There is currently no complete analogue of Kuratowski's theorem for surfaces of genus g, although N. Robertson and P. Seymour have proved that there exists a finite collection of 'forbidden' subgraphs of genus g, for each value of g, corresponding to the forbidden subgraphs K_5 and $K_{3,3}$ for graphs of genus 0.

In the case of Euler's formula we are more fortunate, since there is a natural generalization for graphs of genus g. In this generalization, a **face** of a graph of genus g is defined in the obvious way.

THEOREM 14.2. Let G be a connected graph of genus g with n vertices, m edges and f faces. Then $n - m + f = 2 - 2g$.

Sketch of proof. We outline the main steps in the proof, omitting the details.

Without loss of generality, we may assume that G is drawn on the surface of a sphere with g handles. We can also assume that the curves A at which the handles meet the sphere are cycles of G, by shrinking those cycles that contain these curves in their interior.

We next disconnect each handle at one end, in such a way that the handle has a free end E and the sphere has a corresponding hole H. We may assume that the cycle corresponding to the end of the handle appears at both the free end E and at the other end, since the additional vertices and edges required for this exactly balance each other, leaving $n - m + f$ unchanged (see Fig. 14.2).

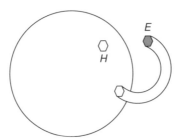

Fig. 14.2

We complete the proof by telescoping each of these handles, leaving a sphere with $2g$ holes in it. This telescoping process does not change the value of $n - m + f$. But for a sphere, $n - m + f = 2$, and hence for a sphere with $2g$ holes in it, $n - m + f = 2 - 2g$. The result follows immediately. //

COROLLARY 14.3. *The genus $g(G)$ of a simple graph G with n (≥ 4) vertices and m edges satisfies the inequality*
$$g(G) \geq \lceil (m - 3n)/6 + 1 \rceil.$$

Proof. Since each face is bounded by at least three edges, we have (as in the proof of Corollary 13.4(i)) $3f \leq 2m$. The result follows on substituting this inequality into Theorem 14.2, and using the fact that the genus of a graph is an integer. //

Just as for the thickness of a graph, little is known about the genus of an arbitrary graph. The usual approach is to use Corollary 14.3 to obtain a lower bound for the genus, and then to try to obtain the required drawing by direct construction.

One case of particular historical importance is that of the genus of the complete graphs. Corollary 14.3 tells us that the genus of K_n satisfies

$$g(K_n) \geq \lceil \{n(n-1)/2 - 3n\}/6 + 1 \rceil$$

or, after a little algebraic manipulation,

$$g(K_n) \geq \lfloor (n-3)(n-4)/12 \rfloor .$$

Percy Heawood asserted in 1890 that the inequality just obtained is an equality, and this was proved in 1968 by Ringel and Youngs after a long and difficult struggle.

THEOREM 14.4 (Ringel and Youngs, 1968). $g(K_n) = \lceil (n-3)(n-4)/12 \rceil$.

Remark. We do not prove this here; see Ringel [35] for a discussion and proof of this theorem.

Further results concerning the drawing of graphs on these surfaces, as well as on 'non-orientable' surfaces (such as the projective plane and the Möbius strip), can be found in Beineke and Wilson [27] or Gross and Tucker [29].

Exercises 14

14.1[s] The surface of a torus can be regarded as a rectangle in which opposite edges are identified (see Fig. 14.3). Using this representation, find drawings of K_5 and $K_{3,3}$ on the torus.

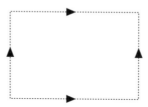

Fig. 14.3

14.2 Using the representation of Exercise 14.1, show that the Petersen graph has genus 1.

14.3[s] (i) Calculate $g(K_7)$ and $g(K_{11})$.
          (ii) Give an example of a complete graph of genus 2.

14.4 (i) Use Theorem 14.4 to prove that there is no value of n for which $g(K_n) = 7$.
        (ii) What is the next integer that is not the genus of any complete graph?

14.5[s] (i) Give an example of a plane graph that is regular of degree 4 and each face of which is a triangle.
          (ii) Show that there is no graph of genus $g \geq 1$ with these properties.

14.6 (i) Obtain a lower bound, analogous to that of Corollary 14.3, for a graph containing no triangles.
        (ii) Deduce that $g(K_{r,s}) \geq \lceil (r-2)(s-2)/4 \rceil$.
             (Ringel has shown that this is an equality.)

14.7* (i) Let G be a non-planar graph that can be drawn on a Möbius strip. Prove that, with the usual notation, $n - m + f = 1$.

(ii) Show how K_5 and $K_{3,3}$ can be drawn on the surface of a Möbius strip.

15 Dual graphs

In Theorems 12.2 and 12.3 we gave necessary and sufficient conditions for a graph to be planar – namely, that it contains no subgraph homeomorphic or contractible to K_5 or $K_{3,3}$. We now discuss conditions of a different kind, involving the concept of duality.

Given a plane drawing of a planar graph G, we construct another graph G^*, called the (**geometric**) **dual** of G. The construction is in two stages:

(i) inside each face f of G we choose a point v^* – these points are the vertices of G^*;

(ii) corresponding to each edge e of G we draw a line e^* that crosses e (but no other edge of G), and joins the vertices v^* in the faces f adjoining e – these lines are the edges of G^*.

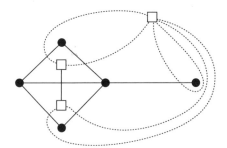

Fig. 15.1

This procedure is illustrated in Fig. 15.1. The vertices v^* of G^* are represented by small squares, the edges e of G by solid lines, and the edges e^* of G^* by dotted lines. Note that an end-vertex or a bridge of G gives rise to a loop of G^*, and that if two faces of G have more than one edge in common, then G^* has multiple edges.

The geometric idea of duality is very old. For example, the 'fifteenth book of Euclid', written about AD 500–600, remarks that the dual of a cube is an octahedron, and that the dual of a dodecahedron is an icosahedron (see Exercise 15.2). Note that any two graphs formed from G in this way must be isomorphic; this is why we called G^* '*the* dual of G' instead of '*a* dual of G'. On the other hand, if G is isomorphic to H, it does not necessarily follow that G^* is isomorphic to H^*; an example that demonstrates this is given in Exercise 15.5.

If G is both plane and connected, then G^* is plane and connected, and there are simple relations between the numbers of vertices, edges and faces of G and G^*.

> **LEMMA 15.1.** *Let G be a plane connected graph with n vertices, m edges and f faces, and let its geometric dual G^* have n^* vertices, m^* edges and f^* faces. Then $n^* = f$, $m^* = m$ and $f^* = n$.*

Proof. The first two relations are direct consequences of the definition of G^*. The third relation follows on substituting these two relations into Euler's formula, applied to both G and G^*. //

Since the dual G^* of a plane graph G is also a plane graph, we can repeat the above construction to form the dual G^{**} of G^*. If G is connected, then the relationship between G^{**} and G is particularly simple, as we now show.

THEOREM 15.2. *If G is a plane connected graph, then G^{**} is isomorphic to G.*

Proof. The result follows immediately, since the construction that gives rise to G^* from G can be reversed to give G from G^*; for example, in Fig. 15.1 the graph G is the dual of the graph G^*. We need to check only that a face of G^* cannot contain more than one vertex of G (it certainly contains at least one), and this follows immediately from the relations $n^{**} = f^* = n$, where n^{**} is the number of vertices of G^{**}. //

If G is a planar graph, then a dual of G can be defined by taking any plane drawing and forming its geometric dual, but uniqueness does not always hold. Since duals have been defined only for planar graphs, it is trivially true to say that a graph is planar if and only if it has a dual. On the other hand, we cannot tell from the above whether a given graph is planar. It is obviously desirable to find a definition of duality that generalizes the geometric dual and enables us in principle to determine whether a given graph is planar. One such definition exploits the relationship under duality between the cycles and cutsets of a planar graph G. We now describe this relationship and use it to obtain the definition we seek. An alternative definition is given in Exercise 15.11.

THEOREM 15.3. *Let G be a planar graph and let G^* be a geometric dual of G. Then a set of edges in G forms a cycle in G if and only if the corresponding set of edges of G^* forms a cutset in G^*.*

Proof. We can assume that G is a connected plane graph. If C is a cycle in G, then C encloses one or more finite faces of C, and thus contains in its interior a non-empty set S of vertices of G^*. It follows immediately that those edges of G^* that cross the edges of C form a cutset of G^* whose removal disconnects G^* into two subgraphs, one with vertex set S and the other containing those vertices that do not lie in S (see Fig. 15.2). The converse implication is similar, and is omitted. //

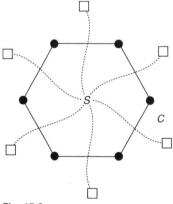

Fig. 15.2

COROLLARY 15.4. *A set of edges of G forms a cutset in G if and only if the corresponding set of edges of G* forms a cycle in G*.*

Proof. The result follows on applying Theorem 15.3 to G* and using Theorem 15.2. //

Using Theorem 15.3 as motivation, we can now give an abstract definition of duality. Note that this definition does not invoke any special properties of planar graphs, but concerns only the relationship between two graphs.

We say that a graph G* is an **abstract dual** of a graph G if there is a one–one correspondence between the edges of G and those of G*, with the property that a set of edges of G forms a cycle in G if and only if the corresponding set of edges of G* forms a cutset in G*. For example, Fig. 15.3 shows a graph and its abstract dual, with corresponding edges sharing the same letter.

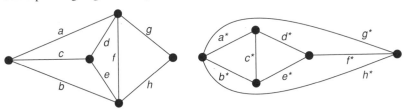

Fig. 15.3

It follows from Theorem 15.3 that the concept of an abstract dual generalizes that of a geometric dual, in the sense that if G is a plane graph and G* is a geometric dual of G, then G* is an abstract dual of G. We should like to obtain analogues for abstract duals of some results on geometric duals. We present just one of these here – the analogue for abstract duals of Theorem 15.2.

THEOREM 15.5. *If G* is an abstract dual of G, then G is an abstract dual of G*.*

Remark. Note that we do not require that G should be connected.

Proof. Let C be a cutset of G and let C* denote the corresponding set of edges of G*. We show that C* is a cycle of G*. By the first part of Exercise 5.12, C has an even number of edges in common with any cycle of G, and so C* has an even number of edges in common with any cutset of G*. It follows from the second part of Exercise 5.12 that C* is either a cycle in G* or an edge-disjoint union of at least two cycles. But the second possibility cannot occur, since we can show similarly that cycles in C* correspond to edge-disjoint unions of cutsets in G, and so C would be an edge-disjoint union of at least two cutsets, rather than a single cutset. //

Although the definition of an abstract dual may seem strange, it turns out to have the properties required of it. We saw in Theorem 15.3 that a planar graph has an abstract dual (for example, any geometric dual). We now show that the converse result is true – that any graph with an abstract dual must be planar. This gives us an abstract definition of duality that generalizes the geometric dual and characterizes planar graphs. It turns out that the definition of an abstract dual is a natural consequence of the study of duality in matroid theory (see Section 32).

THEOREM 15.6. *A graph is planar if and only if it has an abstract dual.*

Remark. There are several proofs of this result. We outline a proof that uses Kuratowski's theorem.

Sketch of proof. As mentioned above, it is sufficient to prove that if G is a graph with an abstract dual G^*, then G is planar. The proof is in four steps.

(i) We note first that if an edge e is removed from G, then the abstract dual of the remaining graph may be obtained from G^* by contracting the corresponding edge e^*. On repeating this procedure, we deduce that, if G has an abstract dual, then so does any subgraph of G.

(ii) We next observe that if G has an abstract dual, and G' is homeomorphic to G, then G' also has an abstract dual. This follows from the fact that the insertion or removal in G of a vertex of degree 2 results in the addition or deletion of a 'multiple edge' in G^*.

(iii) The third step is to show that neither K_5 nor $K_{3,3}$ has an abstract dual. If G^* is a dual of $K_{3,3}$, then since $K_{3,3}$ contains only cycles of length 4 or 6 and no cutsets with two edges, G^* contains no multiple edges and each vertex of G^* has degree at least 4. Hence G^* must have at least five vertices, and thus at least $(5 \times 4)/2 = 10$ edges, which is a contradiction. The argument for K_5 is similar, and is omitted.

(iv) Suppose, now, that G is a non-planar graph with an abstract dual G^*. Then, by Kuratowski's theorem, G has a subgraph H homeomorphic to K_5 or $K_{3,3}$. It follows from (i) and (ii) that H, and hence also K_5 or $K_{3,3}$, must have an abstract dual, contradicting (iii). //

Exercises 15

15.1[s] Find the duals of the graphs in Fig. 15.4 and verify Lemma 15.1 for these.

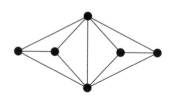

Fig. 15.4

15.2 Show that the dual of the cube graph is the octahedron graph, and that the dual of the dodecahedron graph is the icosahedron graph.

15.3 Show that the dual of a wheel is a wheel.

15.4[s] Use duality to prove that there exists no plane graph with five faces, each pair of which share an edge in common.

15.5[s] Show that the graphs in Fig. 15.5 are isomorphic, but that their geometric duals are not isomorphic.

15.6 (i) Give an example to show that, if G is a disconnected plane graph, then G^{**} is not isomorphic to G.
(ii) Prove the result of part (i) in general.

Fig. 15.5

15.7ˢ Dualize the results of Exercise 13.4.

15.8ˢ Prove that, if G is a 3-connected plane graph, then its geometric dual is a simple graph.

15.9ˢ Let G be a connected plane graph. Using Theorem 5.1 and Corollary 6.3, prove that G is bipartite if and only if its dual G^* is Eulerian.

15.10 (i) Give an example to show that, if G is a connected plane graph, then any spanning tree in G corresponds to the complement of a spanning tree in G^*.
(ii) Prove the result of part (i) in general.

15.11* A graph G^* is a **Whitney dual** of G if there is a one–one correspondence between $E(G)$ and $E(G^*)$ such that, for each subgraph H of G with $V(H) = V(G)$, the corresponding subgraph H^* of G^* satisfies

$$\gamma(H) + \xi(\bar{H}^*) = \xi(G^*),$$

where $\bar{H}^*$ is obtained from G^* by deleting the edges of H^*, and γ and ξ are defined as in Section 9.
(i) Show that this generalizes the idea of a geometric dual.
(ii) Prove that, if G^* is a Whitney dual of G, then G is a Whitney dual of G^*.
(In 1932, H. Whitney proved that a graph is planar if and only if it has such a dual.)

16 Infinite graphs

In this section we show how some of the definitions in previous sections can be extended to infinite graphs. An **infinite graph** G consists of an infinite set $V(G)$ of elements called **vertices**, and an infinite family $E(G)$ of unordered pairs of elements of $V(G)$ called **edges**. If $V(G)$ and $E(G)$ are both countably infinite, then G is a **countable graph**. We exclude the possibility of $V(G)$ being infinite but $E(G)$ finite, as such objects are merely finite graphs together with infinitely many isolated vertices, or of $E(G)$ being infinite but $V(G)$ finite, as such objects are essentially finite graphs but with infinitely many loops or multiple edges.

Many of our earlier definitions ('adjacent', 'incident', 'isomorphic', 'subgraph', 'connected', 'planar', etc.) extend immediately to infinite graphs. The **degree** of a vertex v of an infinite graph is the cardinality of the set of edges incident to v, and may be finite or infinite. An infinite graph, each of whose vertices has finite degree, is **locally finite**; two important examples are the infinite square lattice and the infinite triangular lattice, shown in Figs. 16.1 and 16.2. We similarly define a **locally countable** infinite graph to be one in which each vertex has countable degree.

We can now prove the following simple, but fundamental, result.

THEOREM 16.1. *Every connected locally countable infinite graph is a countable graph.*

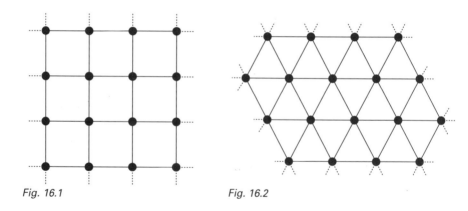

Fig. 16.1   *Fig. 16.2*

Proof. Let v be any vertex of such an infinite graph, and let A_1 be the set of vertices adjacent to v, A_2 be the set of all vertices adjacent to a vertex of A_1, and so on. By hypothesis, A_1 is countable, and hence so are $A_2, A_3, \ldots$, since the union of a countable collection of countable sets is countable. Hence $\{v\}, A_1, A_2, \ldots$ is a sequence of sets whose union is countable and contains every vertex of the infinite graph, by connectedness. The result follows. //

COROLLARY 16.2. *Every connected locally finite infinite graph is a countable graph.*

We can extend the concept of a walk to an infinite graph G. There are essentially three different types of walk in G:

(i)   a *finite walk* is defined exactly as in Section 5;
(ii) a *one-way infinite walk* with initial vertex v_0 is an infinite sequence of edges of the form $v_0v_1, v_1v_2, \ldots$;
(iii) a *two-way infinite walk* is an infinite sequence of edges of the form
      $\ldots, v_{-2}v_{-1}, v_{-1}v_0, v_0v_1, v_1v_2, \ldots$.

One-way and two-way infinite trails and paths are defined analogously, as are the length of a path and the distance between vertices. The following result, known as **König's lemma**, tells us that infinite paths are not difficult to come by.

THEOREM 16.3 (König, 1927). *Let G be a connected locally finite infinite graph. Then, for any vertex v of G, there exists a one-way infinite path with initial vertex v.*

Proof. For each vertex z other than v, there is a non-trivial path from v to z. It follows that there are infinitely many paths in G with initial vertex v. Since the degree of v is finite, infinitely many of these paths must start with the same edge. If vv_1 is such an edge, then we can repeat this procedure for the vertex v_1 and thus obtain a new vertex v_2 and a corresponding edge v_1v_2. By carrying on in this way, we obtain the one-way infinite path $v \to v_1 \to v_2 \to \cdots$. //

The importance of König's lemma is that it allows us to deduce results about infinite graphs from the corresponding results for finite graphs. The following theorem may be regarded as a typical example.

THEOREM 16.4. *If G be a countable graph, every finite subgraph of which is planar, then G is planar.*

Proof. Since G is countable, its vertices may be listed as $v_1, v_2, v_3, \ldots$. We now construct a strictly increasing sequence $G_1 \subset G_2 \subset G_3 \subset \cdots$ of subgraphs of G, by taking G_k to be the subgraph whose vertices are $v_1, \ldots, v_k$ and whose edges are those edges of G joining two of these vertices. Since G_i can be drawn in the plane in only a finite number $m(i)$ of topologically distinct ways, we can construct another infinite graph H whose vertices w_{ij} ($i \geq 1$, $1 \leq j \leq m(i)$) correspond to the drawings of the graphs G_i, and whose edges join those vertices w_{ij} and w_{kl} for which $k = i + 1$ and the plane drawing corresponding to w_{kl} extends the drawing corresponding to w_{ij}. Since H is clearly connected and locally finite, it follows from König's lemma that H contains a one-way infinite path. Since G is countable, this infinite path gives the required plane drawing of G. //

If we assume further axioms of set theory, such as the uncountable version of the axiom of choice, then various results such as the one just proved can be extended to infinite graphs that are not necessarily countable.

We conclude this digression on infinite graphs with a brief discussion on infinite Eulerian graphs. It seems natural to call a connected infinite graph G **Eulerian** if there exists a two-way infinite trail that includes every edge of G; such an infinite trail is a two-way **Eulerian trail**. Note that these definitions require G to be countable. The following theorems give further necessary conditions for an infinite graph to be Eulerian.

THEOREM 16.5. *Let G be a connected countable graph which is Eulerian. Then*
  (i) *G has no vertices of odd degree;*
  (ii) *for each finite subgraph H of G, the infinite graph H obtained by deleting from G the edges of H has at most two infinite connected components;*
  (iii) *if, in addition, each vertex of H has even degree, then H has exactly one infinite connected component.*

Proof. (i) Suppose that P is an Eulerian trail. Then, by the argument given in the proof of Theorem 6.2, each vertex of G must have either even or infinite degree.
  (ii) Let P be split up into three subtrails P_-, P_0 and P_+ in such a way that P_0 is a finite trail containing every edge of H, and possibly other edges as well, and P_- and P_+ are one-way infinite trails. Then the infinite graph K formed by the edges of P_- and P_+, and the vertices incident to them, has at most two infinite components. Since H is obtained by adding only a finite set of edges to K, the result follows.
  (iii) Let the initial and final vertices of P_0 be v and w. We wish to show that v and w are connected in H. If $v = w$, this is obvious. If not, then the result follows on applying Corollary 6.4 to the graph obtained by removing the

edges of H from P_0, this graph having exactly two vertices (v and w) of odd degree, by hypothesis. //

The conditions given in the previous theorem are not only necessary but also sufficient. We state this result formally in the following theorem. Its proof lies may be found in Ore [10].

THEOREM 16.6. *If G is a connected countable graph, then G is Eulerian if and only if the conditions (i), (ii) and (iii) of Theorem 16.5 are satisfied.*

Exercises 16

16.1[s] Give an example of each of the following:
  (i)   an infinite graph with infinitely many end-vertices;
  (ii) an infinite graph with uncountably many vertices and edges;
  (iii) an infinite connected cubic graph;
  (iv) an infinite bipartite graph;
  (v)   an infinite non-planar graph;
  (vi) an infinite tree.

16.2[s] Give an example to show that the conclusion of König's lemma is false if we omit the condition that the infinite graph is locally finite. .

16.3 Show that an infinite graph G can be drawn in Euclidean 3-space if $V(G)$ and $E(G)$ can each be put in one–one correspondence with a subset of the set of real numbers.

16.4* (i)   Find an Eulerian trail in the infinite square lattice S.
  (ii) Verify that S satisfies the conditions of Theorem 16.5.

16.5* Repeat Exercise 16.4 for the infinite triangular lattice.

16.6* Show that the infinite square lattice has both one-way and two-way infinite paths passing exactly once through each vertex.

Chapter 6

Colouring graphs

With colours fairer painted their foul ends.
William Shakespeare (The Tempest)

In this chapter we investigate the colouring of graphs and maps, with special reference to the four-colour theorem and related topics. In Sections 17 and 18, we colour the vertices of a graph so that each edge joins vertices of different colours. Section 19 describes the connection between these vertex colourings and the colouring of maps, and in Section 20 these are related to colouring the edges of a graph. All this material is qualitative, asking *whether* graphs can be coloured with a given number of colours. In Section 21, on chromatic polynomials, we discuss *in how many ways* the colouring can be done.

17 Colouring vertices

If G is a graph without loops, then G is **k-colourable** if we can assign one of k colours to each vertex so that adjacent vertices have different colours. If G is k-colourable, but not $(k-1)$-colourable, we say that G is **k-chromatic**, or that the **chromatic number** of G is k, and write $\chi(G) = k$. For example, Fig. 17.1 shows a graph G for which $\chi(G) = 4$; the colours are denoted by Greek letters. It is thus k-colourable if $k \geq 4$. We shall assume that all graphs here are simple, as multiple edges are irrelevant to our discussion. We shall also assume that they are connected.

It is clear that $\chi(K_n) = n$, and so there are graphs with arbitrarily high chromatic number. At the other end of the scale, $\chi(G) = 1$ if and only if G is a null graph, and

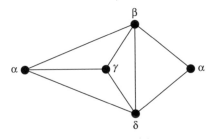

Fig. 17.1

$\chi(G) = 2$ if and only if G is a non-null bipartite graph. Note that every tree is 2-colourable, as is any cycle graph with an even number of vertices.

It is not known which graphs are 3-chromatic, although it is easy to give examples of such graphs. These examples include the cycle graphs or wheels with an odd number of vertices and the Petersen graph. The wheels with an even number of vertices are 4-chromatic.

There is little that we can say about the chromatic number of an arbitrary graph. If the graph has n vertices, then its chromatic number cannot exceed n, and if the graph contains K_r as a subgraph, then its chromatic number cannot be less than r, but these results do not take us very far in general. If, however, we know the degree of each vertex, then we can make progress.

THEOREM 17.1. *If G is a simple graph with largest vertex-degree Δ, then G is $(\Delta+1)$-colourable.*

Proof. The proof is by induction on the number of vertices of G. Let G be a simple graph with n vertices. If we delete any vertex v and its incident edges, then the graph that remains is a simple graph with $n - 1$ vertices and largest vertex-degree at most Δ (see Fig. 17.2). By our induction hypothesis, this graph is $(\Delta+1)$-colourable. A $(\Delta+1)$-colouring for G is then obtained by colouring v with a different colour from the (at most Δ) vertices adjacent to v. //

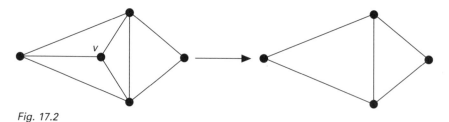

Fig. 17.2

By more careful treatment we can strengthen this theorem a little to give the following result, known as **Brooks' theorem**. Its proof is given in the next section.

THEOREM 17.2 (Brooks, 1941). *If G is a simple connected graph which is not a complete graph, and if the largest vertex-degree of G is Δ (≥ 3), then G is Δ-colourable.*

Both of these theorems are useful if all the vertex-degrees are approximately the same. For example, by Theorem 17.1, every cubic graph is 4-colourable, and by Theorem 17.2, every connected cubic graph, other than K_4, is 3-colourable. On the other hand, if the graph has a few vertices of large degree, then these theorems tell us very little. This is well illustrated by the graph $K_{1,s}$; Brooks' theorem asserts that this is s-colourable, but it is in fact 2-colourable for any s. There is no effective way of avoiding this situation, although there are techniques that help a little.

This rather depressing situation does not arise if we restrict our attention to planar graphs. In fact, we can easily prove the rather strong result that every simple planar graph is 6-colourable.

THEOREM 17.3. *Every simple planar graph is 6-colourable.*

Proof. The proof is similar to that of Theorem 17.1. We prove the theorem by induction on the number of vertices, the result being trivial for simple planar graphs with at most six vertices. Suppose then that G is a simple planar graph with n vertices, and that all simple planar graphs with $n - 1$ vertices are 6-colourable. By Theorem 13.6, G contains a vertex v of degree at most 5. If we delete v and its incident edges, then the graph that remains has $n - 1$ vertices and is thus 6-colourable (see Fig. 17.3). A 6-colouring of G is then obtained by colouring v with a colour different from the (at most five) vertices adjacent to v. //

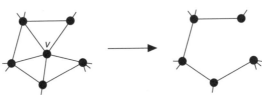

Fig. 17.3

As with Theorem 17.1, this result can be strengthened by more careful treatment. The result is called the **five-colour theorem**.

THEOREM 17.4. *Every simple planar graph is 5-colourable.*

Proof. The proof is similar to that of Theorem 17.3, although the details are more complicated. We prove the theorem by induction on the number of vertices, the result being trivial for simple planar graphs with fewer than six vertices. Suppose then that G is a simple planar graph with n vertices, and that all simple planar graphs with $n - 1$ vertices are 5-colourable. By Theorem 13.6, G contains a vertex v of degree at most 5. As before, the deletion of v leaves a graph with $n - 1$ vertices, which is thus 5-colourable. Our aim is to colour v with one of the five colours, so completing the 5-colouring of G.

If $\deg(v) < 5$, then v can be coloured with any colour not assumed by the (at most four) vertices adjacent to v, completing the proof in this case. We thus suppose that $\deg(v) = 5$, and that the vertices $v_1, \ldots, v_5$ adjacent to v are arranged around v in clockwise order as in Fig. 17.4. If the vertices $v_1, \ldots, v_5$ are all mutually adjacent, then G contains the non-planar graph K_5 as a subgraph, which is impossible. So at least two of the vertices v_i (say, v_1 and v_3) are not adjacent.

We now contract the two edges vv_1 and vv_3. The resulting graph is a planar graph with fewer than n vertices, and is thus 5-colourable. We now reinstate the two edges, giving both v_1 and v_3 the colour originally assigned to v. A 5-colouring of G is then obtained by colouring v with a colour different from the (at most four) colours assigned to the vertices v_i. //

It is natural to ask whether this result can be strengthened further, and this leads to

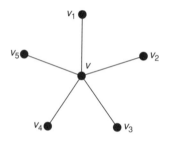

Fig. 17.4

what was formerly one of the most famous unsolved problems in mathematics – the 'four-colour problem'. This problem, in an alternative formulation (see Section 19), was first posed in 1852, and was eventually settled by K. Appel and W. Haken in 1976.

THEOREM 17.5 (Appel and Haken, 1976). *Every simple planar graph is 4-colourable.*

Their proof, which took them several years and a substantial amount of computer time, ultimately depends on a complicated extension of the ideas in the proof of the five-colour theorem. Further information about this proof can be found in Saaty and Kainen [36], or in Beineke and Wilson [27].

We conclude this section with a simple application of vertex colourings. Suppose that a chemist wishes to store five chemicals *a, b, c, d* and *e* in various areas of a warehouse. Some of these chemicals react violently when in contact, and so must be kept in separate areas. In the following table, an asterisk indicates those pairs of chemicals that must be separated. How many areas are needed?

	a	*b*	*c*	*d*	*e*
a	–	*	*	*	–
b	*	–	*	*	*
c	*	*	–	*	–
d	*	*	*	–	*
e	–	*	–	*	–

To answer this, we draw the graph whose vertices correspond to the five chemicals, with two vertices adjacent whenever the corresponding chemicals are to be kept apart (see Fig. 17.5). If we now colour the vertices, as shown by the Greek letters, then the

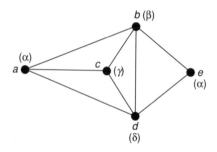

Fig. 17.5

colours correspond to the areas needed. In this case, the chromatic number is 4, and so four areas are needed. For example, chemicals *a* and *e* can be stored in area α, and chemicals *b*, *c* and *d* can be stored in areas β, γ and δ, respectively.

Exercises 17

17.1[s] Find the chromatic number of each graph in Fig 17.6.

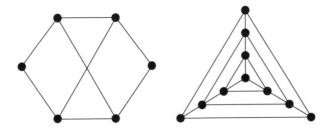

Fig. 17.6

17.2 Find the chromatic number of each graph in Fig. 17.7.

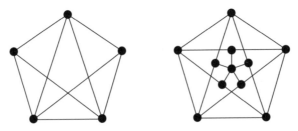

Fig. 17.7

17.3[s] In the table of Fig. 2.9, locate all the 2-chromatic, 3-chromatic and 4-chromatic graphs.

17.4 What is the chromatic number of
  (i)   each of the Platonic graphs?
  (ii) the complete tripartite graph $K_{r,s,t}$?
  (iii) the *k*-cube Q_k?

17.5[s] Compare the upper bound for the chromatic number given by Brooks' theorem with the correct value, for
  (i)   the Petersen graph;
  (ii) the *k*-cube Q_k.

17.6 A lecture timetable is to be drawn up. Since some students wish to attend several lectures, certain lectures must not coincide. The asterisks in the following table show which pairs of lectures cannot coincide. How many periods are needed to timetable all seven lectures?

	a	*b*	*c*	*d*	*e*	*f*	*g*
a	—	*	*	*	—	—	*
b	*	—	*	*	*	—	*
c	*	*	—	*	—	*	—
d	*	*	*	—	—	*	—
e	—	*	—	—	—	—	—
f	—	—	*	*	—	—	*
g	*	*	—	—	—	*	—

17.7^s   Let G be a simple graph with n vertices, which is regular of degree d. By considering the number of vertices that can be assigned the same colour, prove that $\chi(G) \geq n/(n-d)$.

17.8   Let G be a simple planar graph containing no triangles.
   (i)   Using Euler's formula, show that G contains a vertex of degree at most 3.
   (ii)   Use induction to deduce that G is 4-colourable.
   (In fact, it can be proved that G is 3-colourable.)

17.9*   Generalize the results of the previous exercise to the cases where
   (i) G has girth r;
   (ii) G has thickness t.

17.10*   Try to prove the four-colour theorem by adapting the above proof of the five-colour theorem. At what point does the proof fail?

17.11*   A graph G is **k-critical** if $\chi(G) = k$ and if the deletion of any vertex yields a graph with smaller chromatic number.
   (i)   Find all 2-critical and 3-critical graphs.
   (ii)   Give an example of a 4-critical graph.
   (iii)   Prove that, if G is k-critical, then
      (a)   every vertex of G has degree at least $k-1$;
      (b) G has no cut-vertices.

17.12*   Let G be a countable graph, each finite subgraph of which is k-colourable.
   (i)   Use König's lemma to prove that G is k-colourable.
   (ii)   Deduce that every countable planar graph is 4-colourable.

18   Brooks' theorem

In order to avoid disturbing the continuity, we deferred the proof of Brooks' theorem (Theorem 17.2). This proof will now be given.

> **THEOREM 17.2** *If G is a simple connected graph which is not a complete graph, and if the largest vertex degree of G is Δ (≥ 3), then G is Δ-colourable.*

Proof. The proof is by induction on the number of vertices of G. Suppose that G has n vertices. If any vertex of G has degree less than Δ, then we can complete the proof by imitating the proof of Theorem 17.1. We may thus suppose that G is regular of degree Δ.

If we delete a vertex v and its incident edges, then the graph that remains has $n-1$ vertices and the largest vertex degree is still at most Δ. By our induction hypothesis, this graph is Δ-colourable. Our aim is now to colour v with one of the Δ colours. We can suppose that the vertices $v_1, \ldots, v_\Delta$ adjacent to v are arranged around v in clockwise order, and that they are coloured with distinct colours $c_1, \ldots, c_\Delta$, since otherwise there would be a spare colour that could be used to colour v.

We now define H_{ij} ($i \neq j$, $1 \leq i, j \leq \Delta$) to be the subgraph of G whose vertices are those coloured c_i or c_j and whose edges are those joining a vertex coloured c_i and a vertex coloured c_j. If the vertices v_i and v_j lie in different components of H_{ij}, then we can interchange the colours of all the vertices in the component of H_{ij} containing v_i

(see Fig. 18.1). The result of this recolouring is that v_i and v_j both have colour c_j, enabling v to be coloured c_i. We may thus assume that, given any i and j, v_i and v_j are connected by a path that lies entirely in H_{ij}. We denote the component of H_{ij} containing v_i and v_j by C_{ij}.

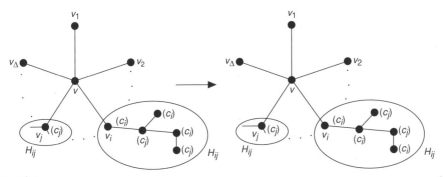

Fig. 18.1

If v_i is adjacent to more than one vertex with colour c_j, then there is a colour (other than c_i) that is not assumed by any vertex adjacent to v_i. In this case, v_i can be recoloured using this colour, enabling v to be coloured with colour c_i. If this does not happen, then we can use a similar argument to show that every vertex of C_{ij} (other than v_i and v_j) must have degree 2. For, if w is the first vertex of the path from v_i to v_j with degree greater than 2, then w can be recoloured with a colour different from c_i or c_j, thereby destroying the property that v_i and v_j are connected by a path lying entirely in C_{ij} (see Fig. 18.2). We can thus assume that, for any i and j, the component C_{ij} consists only of a path from v_i to v_j.

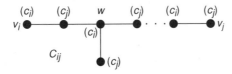

Fig. 18.2

We can also assume that two paths of the form C_{ij} and C_{jl} (where $i \neq l$) intersect only at v_j, since any other point of intersection x can be recoloured with a colour different from c_i, c_j or c_l (see Fig. 18.3), contradicting the fact that v_i and v_j are connected by a path.

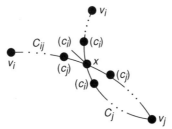

Fig. 18.3

To complete the proof, we choose two vertices v_i and v_j that are not adjacent, and let y be the vertex with colour c_j that is adjacent to v_i. If C_{il} is a path (for some $l \neq j$), then we can interchange the colours of the vertices in this path without affecting the colouring of the rest of the graph (see Fig. 18.4). But if we carry out this interchange, then y would be a vertex common to the paths C_{ij} and C_{jl}, which is a contradiction. This contradiction establishes the theorem. //

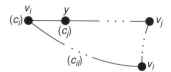

Fig. 18.4

19   Colouring maps

The four-colour problem arose historically in connection with the colouring of maps. Given a map containing several countries, we may ask how many colours are needed to colour them so that no two countries with a boundary line in common share the same colour. Probably the most familiar form of the four-colour theorem is the statement that every map can be coloured with only four colours. For example, Fig. 19.1 shows a map that has been coloured with four colours.

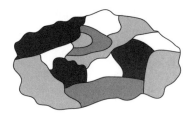

Fig. 19.1

In order to make this statement precise, we must explain what we mean by a 'map'. Since the two colours on either side of an edge must be different, we need to exclude maps containing a bridge (see Fig. 19.2). We also exclude vertices of degree 2, as they can easily be eliminated (see Fig. 19.3). To cover these, and similar cases, we define a **map** to be a 3-connected plane graph; thus a map contains no cutsets with 1 or 2 edges, and in particular no vertices of degree 1 or 2. As we shall see, the exclusion of bridges here corresponds to the exclusion of loops in Section 17.

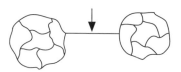

Fig. 19.2   *Fig. 19.3*

We now define a map to be **k-colourable(f)** if its faces can be coloured with k colours so that no two faces with a boundary edge in common have the same colour. To avoid confusion, we use **k-colourable(v)** to mean k-colourable in the usual sense. For example, the map in Fig. 19.4 is 3-colourable(f) and 4-colourable(v).

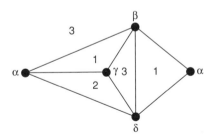

Fig. 19.4

The four-colour theorem for maps is thus the assertion that every map is 4-colourable(f). In Corollary 19.3 we prove the equivalence of the two forms of the four-colour theorem. In the meantime, we investigate the conditions under which a map can be coloured with two colours. These conditions take a particularly simple form.

THEOREM 19.1. *A map G is 2-colourable(f) if and only if G is an Eulerian graph.*

First proof. ⟹ For each vertex v of G, the faces surrounding v must be even in number, since they can be coloured with two colours. It follows that each vertex has even degree and so, by Theorem 6.2, G is Eulerian.

⟸ If G is Eulerian, we colour its faces in two colours as follows. Choose any face F and colour it red. Draw a curve from a point x in F to a point in each other face, passing through no vertex of G. If such a curve crosses an even number of edges, colour the face red; otherwise, colour it blue (see Fig. 19.5). This colouring is well defined, as can be seen by taking a 'cycle' of two such curves and proving that it crosses an even number of edges of G, using the fact that each vertex has an even number of edges incident with it. //

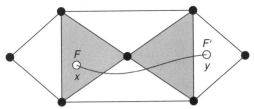

Fig. 19.5

A simpler proof of Theorem 19.1 involves translating the problem into one of colouring the vertices of the dual graph. We first justify this procedure, and then illustrate it by giving our alternative proof of Theorem 19.1 and by proving the equivalence of the two forms of the four-colour theorem.

THEOREM 19.2. *Let G be a plane graph without loops, and let G* be a geometric dual of G. Then G is k-colourable(v) if and only if G* is k-colourable(f).*

Proof. ⇒ We can assume that G is simple and connected, so that G^* is a map. If we have a k-colouring(v) for G, then we can k-colour the faces of G^* so that each face inherits the colour of the unique vertex that it contains (see Fig. 19.6). No two adjacent faces of G^* can have the same colour because the vertices of G that they contain are adjacent in G and so are differently coloured. Thus G^* is k-colourable(f).

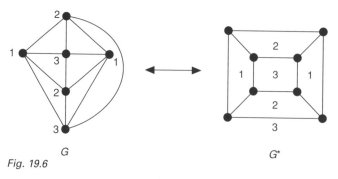

Fig. 19.6

⇐ Suppose now that we have a k-colouring(f) of G^*. Then we can k-colour the vertices of G so that each vertex inherits the colour of the face containing it. No two adjacent vertices of G have the same colour, by reasoning similar to the above. Thus G is k-colourable(v). //

It follows that we can dualize any theorem on the colouring of the vertices of a planar graph to give a theorem on the colouring of the faces of a map, and conversely. As an example of this, consider Theorem 19.1.

THEOREM 19.1. *A map G is 2-colourable(f) if and only if G is an Eulerian graph.*

Second proof. By Exercise 15.9, the dual of an Eulerian planar graph is a bipartite planar graph, and conversely. It is therefore sufficient to note that a connected planar graph without loops is 2-colourable(v) if and only if it is bipartite. //

We can similarly prove the equivalence of the two forms of the four-colour theorem.

COROLLARY 19.3. *The four-colour theorem for maps is equivalent to the four-colour theorem for planar graphs.*

Proof. ⇒ We may assume that G is a simple connected plane graph. Then its geometric dual G^* is a map, and the 4-colourability(v) of G follows immediately from the fact that this map is 4-colourable(f), by Theorem 19.2.
⇐ Conversely, let G be a map and let G^* be its geometric dual. Then G^* is a simple planar graph and is therefore 4-colourable(v). It follows immediately that G is 4-colourable(f). //

Duality can also be used to prove the following theorem.

THEOREM 19.4. *Let G be a cubic map. Then G is 3-colourable(f) if and only if each face is bounded by an even number of edges.*

Proof. ⇒ Given any face F of G, the faces of G that surround F must alternate in colour. Thus there must be an even number of them, and so each face is bounded by an even number of edges (see Fig. 19.7).

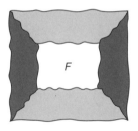

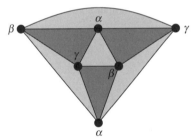

Fig. 19.7   Fig. 19.8

⇐ We prove the dual result, that if G is a simple connected plane graph in which each face is a triangle and each vertex has even degree (that is, G is Eulerian), then G is 3-colourable(v). We shall denote the three colours by α, β and γ.

Since G is Eulerian, it follows from Theorem 19.1 that its faces can be coloured with two colours, red and blue. The required 3-colouring of the vertices of G is then obtained by colouring the vertices of any red face so that the colours α, β and γ appear in clockwise order, and colouring the vertices of any blue face so that these colours appear in anti-clockwise order (see Fig. 19.8). This vertex colouring can be extended to the whole graph, thus proving the theorem. //

In the above theorem, we assumed that the map is cubic. This need not be a severe restriction, as the following theorem shows.

> **THEOREM 19.5.** *In order to prove the four-colour theorem, it is sufficient to prove that each cubic map is 4-colourable(f).*

Proof. By Corollary 19.3, it is sufficient to prove that the 4-colourability(f) of every cubic map implies the 4-colourability(f) of any map. Let G be any map. If G has any vertices of degree 2, then we can remove them without affecting the colouring. It remains only to eliminate vertices of degree 4 or more. But if v is such a vertex, then we can stick a 'patch' over v, as in Fig. 19.9. Repeating this for all such vertices, we obtain a cubic map that is 4-colourable(f) by hypothesis. The required 4-colouring of the faces of G is then obtained by shrinking each patch to a single vertex and reinstating each vertex of degree 2. //

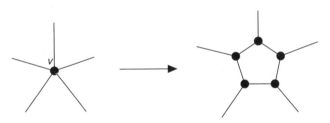

Fig. 19.9

Exercises 19

19.1   Consider the map in Fig. 19.10, in which the countries are to be coloured red, blue, green and yellow.

   (i)   Show that country A must be red.

   (ii)   What colour is country B?

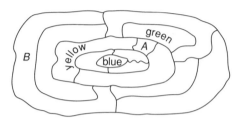

Fig. 19.10

19.2ˢ   Find the minimum number of colours needed to colour the faces of each of the Platonic graphs, so that neighbouring faces are coloured differently.

19.3ˢ   Give an example of a plane graph that is both 2-colourable(f) and 2-colourable(v).

19.4   The plane is divided into a finite number of regions by drawing infinite straight lines in an arbitrary manner. Show that these regions can be 2-coloured.

19.5ˢ   By dualizing the proof of Theorem 17.3, prove the six-colour theorem for maps.

19.6*   By dualizing the proof of Theorem 17.4, prove the five-colour theorem for maps.

19.7*   Let G be a simple plane graph with fewer than 12 faces, and suppose that each vertex of G has degree at least 3.

   (i)   Use Exercise 13.5 to prove that G is 4-colourable(v).

   (ii)   Dualize the result of part (i).

19.8*   (i)   Prove that, if a toroidal graph is embedded on the surface of a torus, then its faces can be coloured with seven colours.

   (ii)   Find a toroidal graph whose faces cannot be coloured with six colours.

20   Colouring edges

In this section we colour the edges of a graph. As we shall see, the four-colour theorem for planar graphs is equivalent to a theorem concerning edge colourings of cubic maps.

A graph G is **k-colourable(e)** (or k-edge colourable) if its edges can be coloured with k colours so that no two adjacent edges have the same colour. If G is k-colourable(e) but not $(k-1)$-colourable(e), we say that the **chromatic index** of G is k, and write $\chi'(G) = k$. For example, Fig. 20.1 shows a graph G for which $\chi'(G) = 4$.

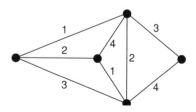

Fig. 20.1

Note that, if Δ is the largest vertex degree of G, then $\chi'(G) \geq \Delta$. The following result, known as **Vizing's theorem**, gives very sharp bounds for the chromatic index of a simple graph G. Its proof can be found in Bondy and Murty [7] or Fiorini and Wilson [28].

THEOREM 20.1 (Vizing, 1964). *If G is a simple graph with largest vertex-degree Δ, then $\Delta \leq \chi'(G) \leq \Delta + 1$.*

It is not known which graphs have chromatic index Δ and which have chromatic index $\Delta + 1$. However, the results for particular types of graphs can easily be found. For example, $\chi'(C_n) = 2$ or 3, depending on whether n is even or odd, and $\chi'(W_n) = n - 1$ if $n \geq 4$.

We now determine the corresponding results for complete graphs.

THEOREM 20.2. $\chi'(K_n) = n$ *if n is odd $(n \neq 1)$, and* $\chi'(K_n) = n - 1$ *if n is even.*

Proof. The result is trivial if $n = 2$. We therefore assume that $n \geq 3$.

If n is odd, then we can n-colour the edges of K_n by placing the vertices of K_n in the form of a regular n-gon, colouring the edges around the boundary with a different colour for each edge, and then colouring each remaining edge with the colour used for the boundary edge parallel to it (see Fig. 20.2). The fact that K_n is not $(n-1)$-colourable(e) follows by observing that the largest possible number of edges of the same colour is $(n-1)/2$, and so K_n has at most $(n-1)/2 \times \chi'(K_n)$ edges.

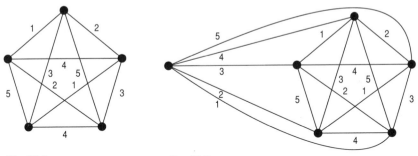

Fig. 20.2   Fig. 20.3

If n is even, then we first obtain K_n by joining the complete graph K_{n-1} to a single vertex. If we now colour the edges of K_{n-1} as above, then there is one colour missing at each vertex, and these missing colours are all different. We complete the edge colouring of K_n by colouring the remaining edges with these missing colours (see Fig. 20.3). //

We now show the connection between the four colour theorem and the colouring of the edges of a graph. This connection accounts for much of the interest in edge colourings.

THEOREM 20.3. *The four-colour theorem is equivalent to the statement that $\chi'(G) = 3$ for each cubic map G.*

Proof. ⇒ Suppose that we have a 4-colouring of the faces of *G*, where the colours are denoted by α = (1, 0), β = (0, 1), γ = (1, 1) and δ = (0, 0). We can then construct a 3-colouring of the edges of *G* by colouring each edge *e* with the colour obtained by adding together (modulo 2) the colours of the two faces adjoining *e*. For example, if *e* adjoins two faces coloured α and γ, then *e* is coloured β, since (1, 0) + (1, 1) = (0, 1). Note that the colour δ cannot occur in this edge colouring, since the two faces adjoining each edge must be distinct. Moreover, no two adjacent edges can share the same colour. We thus have the required edge colouring (see Fig. 20.4).

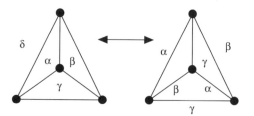

Fig. 20.4

⇐ Suppose now that we have a 3-colouring of the edges of *G*. Then there is an edge of each colour at each vertex. The subgraph determined by those edges coloured α or β is regular of degree 2, and so, by an obvious extension of Theorem 19.1 to disconnected graphs, we can colour its faces with two colours, 0 and 1. In a similar way, we can colour the faces of the subgraph determined by those edges coloured α or γ with the colours 0 and 1. Thus, we can assign to each face of *G* two coordinates (*x*, *y*), where each of *x* and *y* is 0 or 1. Since the coordinates assigned to two adjacent faces of *G* must differ in at least one place, these coordinates (1, 0), (0, 1), (1, 1), (0, 0), give the required 4-colouring of the faces of *G*. //

We conclude this section with a theorem of Dénes König on the chromatic index of a bipartite graph.

> **THEOREM 20.4 (König 1916).** *If G is a bipartite graph with largest vertex-degree* Δ, *then* $\chi'(G) = \Delta$.

Remark. The method of proof is similar to that given in Section 18 – we consider a two-coloured subgraph H_{ij} and interchange the colours.

Proof. We use induction on the number of edges of *G*, and prove that if all but one of the edges have been coloured with at most Δ colours, then there is a Δ-colouring of the edges of *G*.

So suppose that each edge of *G* has been coloured, except for the edge *vw*. Then there is at least one colour missing at the vertex *v*, and at least one colour missing at the vertex *w*. If some colour is missing from both *v* and *w*, then we colour the edge *vw* with this colour. If this is not the case, then let α be a colour missing at *v*, and β be a colour missing at *w*, and let $H_{\alpha\beta}$ be the connected subgraph of *G* consisting of the vertex *v* and those edges and vertices of *G* that can be reached from *v* by a path consisting entirely of edges coloured α or β (see Fig. 20.5).

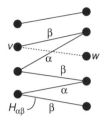

Fig. 20.5

Since G is bipartite, the subgraph $H_{\alpha\beta}$ cannot contain the vertex w, and so we can interchange the colours α and β in this subgraph without affecting w or the rest of the colouring. The edge vw can now be coloured β, thereby completing the colouring of the edges of G. //

COROLLARY 20.5. $\chi'(K_{r,s}) = \max(r, s)$.

Exercises 20

20.1ˢ   Find the chromatic index of each graph in Fig. 20.6.

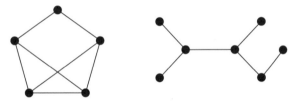

Fig. 20.6

20.2   Find the chromatic index of each graph in Fig. 20.7.

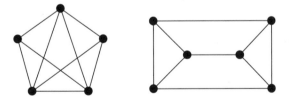

Fig. 20.7

20.3ˢ   In the table of Fig. 2.9, locate all the graphs with chromatic index 2, 3 and 4.

20.4ˢ   Compare the lower and upper bounds for the chromatic index given by Vizing's theorem with the correct value, for
   (i)   the cycle graph C_7;
   (ii) the complete graph K_8;
   (iii) the complete bipartite graph $K_{4,6}$.

20.5   What is the chromatic index of each of the Platonic graphs?

20.6ˢ   By exhibiting an explicit colouring for the edges of $K_{r,s}$, give an alternative proof of Corollary 20.5.

20.7ˢ   Prove that if G is a cubic Hamiltonian graph, then $\chi'(G) = 3$.

20.8   (i)   By considering the possible 3-colourings of the outer 5-cycle, prove that the Petersen graph has chromatic index 4.

   (ii)   Using part (i) and Exercise 20.7, deduce that the Petersen graph is non-Hamiltonian.

20.9*   Let G be a simple graph with an odd number of vertices. Prove that if G is regular of degree Δ, then $\chi'(G) = \Delta + 1$.

20.10*   (i)   Let G be a simple graph which is not a null graph. Prove that $\chi'(G) = \chi(L(G))$, where $L(G)$ is the line graph of G.

   (ii)   By combining part (i) with Brooks' theorem, prove Vizing's theorem in the case $\Delta = 3$.

21   Chromatic polynomials

We conclude this chapter by returning to vertex colourings. In this section we associate with each graph a function that tells us, among other things, whether or not the graph is 4-colourable. By investigating this function, we may hope to gain useful information about the four-colour theorem. Without loss of generality, we restrict our attention to simple graphs.

Let G be a simple graph, and let $P_G(k)$ be the number of ways of colouring the vertices of G with k colours so that no two adjacent vertices have the same colour. P_G is called (for the time being) the **chromatic function** of G. For example, if G is the tree shown in Fig. 21.1, then $P_G(k) = k(k-1)^2$, since the middle vertex can be coloured in k ways, and then the end-vertices can each be coloured in any of $k-1$ ways. This result can be extended to show that, if T is any tree with n vertices, then $P_G(k) = k(k-1)^{n-1}$. Similarly, if G is the complete graph K_3 in Fig. 21.2, then $P_G(k) = k(k-1)(k-2)$. This can be extended to $P_G(k) = k(k-1)(k-2) \cdots (k-n+1)$ if G is the graph K_n.

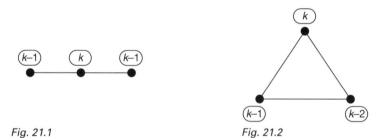

Fig. 21.1   Fig. 21.2

It is clear that *if $k < \chi(G)$, then $P_G(k) = 0$*, and that *if $k \geq \chi(G)$, then $P_G(k) > 0$*. Note that the four-colour theorem is equivalent to the statement: *if G is a simple planar graph, then $P_G(4) > 0$*.

If we are given an arbitrary simple graph, it is usually difficult to obtain its chromatic function by inspection. The following theorem and corollary give us a systematic method for obtaining the chromatic function of a simple graph in terms of the chromatic functions of null graphs.

THEOREM 21.1. *Let G be a simple graph, and let $G - e$ and G/e be the graphs obtained from G by deleting and contracting an edge e. Then*
$$P_G(k) = P_{G-e}(k) - P_{G/e}(k).$$

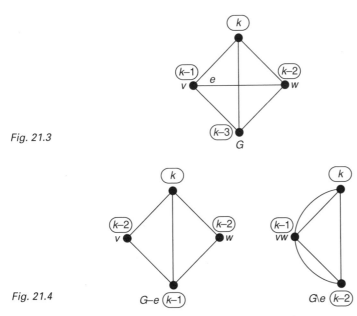

Fig. 21.3

Fig. 21.4

For example, let G be the graph shown in Fig. 21.3. The corresponding graphs $G - e$ and G/e are shown in Fig. 21.4, and the theorem states that

$$k(k-1)(k-2)(k-3) = [k(k-1)(k-2)^2] - [k(k-1)(k-2)].$$

Proof. Let $e = vw$. The number of k-colourings of $G - e$ in which v and w have *different* colours is unchanged if the edge e is drawn joining v and w, and is therefore equal to $P_G(k)$. Similarly, the number of k-colourings of $G - e$ in which v and w have *the same* colour is unchanged if v and w are identified, and is therefore equal to $P_{G/e}(k)$. The total number $P_{G-e}(k)$ of k-colourings of $G - e$ is therefore $P_G(k) + P_{G/e}(k)$, as required. //

COROLLARY 21.2. *The chromatic function of a simple graph is a polynomial.*

Proof. We continue the above procedure by choosing edges in $G - e$ and G/e and deleting and contracting them. We then repeat the procedure for these four new graphs, and so on. The process terminates when no edges remain – in other words, when each remaining graph is a null graph. Since the chromatic function of a null graph is a polynomial ($= k^r$, where r is the number of vertices), it follows by repeated application of Theorem 21.1 that the chromatic function of the graph G must be a sum of polynomials, and so must itself be a polynomial. //

A worked example that illustrates this procedure is given later in the section. In practice, we do not need to reduce each graph to a null graph. It is enough to reduce each graph to graphs whose chromatic functions we already know, such as trees.

In the light of Corollary 21.2, we can now call $P_G(k)$ the **chromatic polynomial** of G. Note from the above proof that, if G has n vertices, then $P_G(k)$ is of degree n, since no new vertices are introduced at any stage. Since the construction yields only one null graph on n vertices, the coefficient of k^n is 1. We can also show (see Exercise 21.6) that the coefficients alternate in sign, and that the coefficient of k^{n-1} is $-m$, where m is the

Fig. 21.5

number of edges of G. Note that we cannot colour a graph if no colours are available, and so the constant term of any chromatic polynomial is 0.

We now give an example to illustrate the above ideas. We use Theorem 21.1 to find the chromatic polynomial of the graph G of Fig. 21.5 and then verify that this polynomial has the form $k^5 - 7k^4 + ak^3 - bk^2 + ck$, where a, b and c are positive constants, as expected from the previous paragraph. It is convenient at each stage to draw the graph itself, rather than write its chromatic polynomial. For example, instead of writing $P_G(k) = P_{G-e}(k) - P_{G\backslash e}(k)$, where G, $G - e$ and $G\backslash e$ are the graphs of Figs. 21.3 and 21.4, we write down the 'equation' in Fig. 21.6.

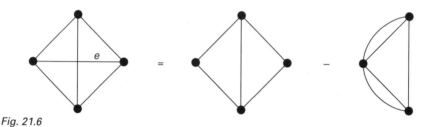

Fig. 21.6

With this convention, and ignoring multiple edges as we proceed, we have

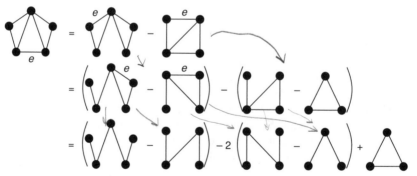

Thus

$$P_G(k) = k(k-1)^4 - 3k(k-1)^3 + 2k(k-1)^2 + k(k-1)(k-2)$$

$$= k^5 - 7k^4 + 18k^3 - 20k^2 + 8k.$$

Note that this result has the required form $k^5 - 7k^4 + ak^3 - bk^2 + ck$, where a, b and c are positive constants.

We conclude this chapter by recalling from Exercise 17.6 how vertex colourings are related to timetabling problems. Suppose that we wish to arrange the times at which certain lectures are to be given. Some pairs of lectures cannot be given at the same time, since there may be students who wish to attend both. In order to construct a timetable, we construct a graph whose vertices correspond to the lectures and whose edges join pairs of lectures that cannot be scheduled at the same time. If we associate a colour with each time available for lectures, then a colouring of the vertices corresponds to a timetabling of the lectures. The chromatic number of the graph tells us the number of lecture periods needed, and the chromatic polynomial tells us how many ways there are of timetabling the lectures.

Exercises 21

21.1ˢ Write down the chromatic polynomials of
(i)   the complete graph K_6;
(ii) the complete bipartite graph $K_{1,5}$.
In how many ways can these graphs be coloured with 7 colours?

21.2 (i) Find the chromatic polynomials of the six connected simple graphs on four vertices.
(ii) Verify that each of the polynomials in part (i) has the form

$$k^4 - mk^3 + ak^2 - bk,$$

where m is the number of edges, and a and b are positive constants.

21.3ˢ Find the chromatic polynomials of
(i)   the complete bipartite graph $K_{2,5}$;
(ii) the cycle graph C_5.

21.4* (i) Prove that the chromatic polynomial of $K_{2,s}$ is

$$k(k-1)^s + k(k-1)(k-2)^s.$$

(ii) Prove that the chromatic polynomial of C_n is

$$(k-1)^n + (-1)^n(k-1).$$

21.5 Prove that, if G is a disconnected simple graph, then its chromatic polynomial $P_G(k)$ is the product of the chromatic polynomials of its components. What can you say about the degree of the lowest non-vanishing term?

21.6* Let G be a simple graph with n vertices and m edges. Use induction on m, together with Theorem 21.1, to prove that
(i)   the coefficient of k^{n-1} is $-m$;
(ii) the coefficients of $P_G(k)$ alternate in sign.

21.7ˢ (i) Use the results of Exercises 21.5 and 21.6 to prove that, if
$P_G(k) = k(k-1)^n$,
then G is a tree on n vertices.
(ii) Find three graphs with chromatic polynomial

$$k^5 - 4k^4 + 6k^3 - 4k^2 + k.$$

Chapter 7

Digraphs

By indirections find directions out.
William Shakespeare (Hamlet)

This chapter and the following one deal with digraphs and their applications. In Section 22 we give some basic definitions, and discuss whether we can 'direct' the edges of a graph so that the resulting digraph is strongly connected. This is followed by a brief discussion of critical path analysis, and, in Section 23, by a discussion of Eulerian and Hamiltonian trails and cycles, with particular reference to tournaments. We conclude the chapter by studying the classification of states of a Markov chain.

22 Definitions

A **directed graph**, or **digraph**, D consists of a non-empty finite set $V(D)$ of elements called **vertices**, and a finite family $A(D)$ of ordered pairs of elements of $V(D)$ called **arcs**. We call $V(D)$ the **vertex set** and $A(D)$ the **arc family** of D. An arc (v, w) is usually abbreviated to vw. Thus in Fig. 22.1, $V(D)$ is the set $\{u, v, w, z\}$ and $A(D)$ consists of the arcs uv, vv, vw (twice), wv, wu and zw, the ordering of the vertices in an arc being indicated by an arrow. If D is a digraph, the graph obtained from D by 'removing the arrows' (that is, by replacing each *arc* of the form vw by a corresponding *edge* vw) is the **underlying graph** of D (see Fig. 22.2).

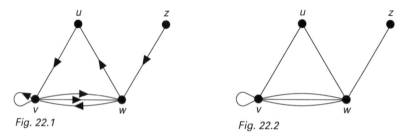

Fig. 22.1   Fig. 22.2

D is a **simple digraph** if the arcs of D are all distinct, and if there are no 'loops' (arcs of the form vv). Note that the underlying graph of a simple digraph need not be a simple graph (see Fig. 22.3).

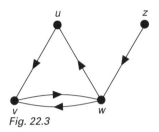

Fig. 22.3

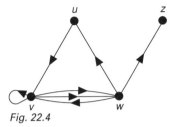

Fig. 22.4

We can imitate many of the definitions given in Section 2 for graphs. For example, two digraphs are **isomorphic** if there is an isomorphism between their underlying graphs that preserves the ordering of the vertices in each arc. Note that the digraphs in Figs. 22.1 and 22.4 are not isomorphic.

Two vertices v and w of a digraph D are **adjacent** if there is an arc in $A(D)$ of the form vw or wv. The vertices v and w are **incident** to such an arc. If D has vertex set $\{v_1, \ldots, v_n\}$, the **adjacency matrix** of D is the $n \times n$ matrix $\mathbf{A} = (a_{ij})$, where a_{ij} is the number of arcs from v_i to v_j.

There are also natural generalizations to digraphs of the definitions of Section 5. A walk in a digraph D is a finite sequence of arcs of the form $v_0v_1, v_1v_2, \ldots, v_{m-1}v_m$. We sometimes write this sequence as $v_0 \to v_1 \to \cdots \to v_m$, and speak of a **walk from v_0 to v_m**. In an analogous way, we can define directed trails, directed paths and directed cycles or, simply, trails, paths and cycles, if there is no possibility of confusion. Note that, although a trail cannot contain a given arc vw more than once, it can contain both vw and wv; for example, in Fig. 22.1, $z \to w \to v \to w \to u$ is a trail.

We can also define connectedness. The two most useful types of connected digraph correspond to whether or not we take account of the direction of the arcs. These definitions are the natural extensions to digraphs of the definitions of connectedness given in Sections 2 and 5.

A digraph D is **connected** if it cannot be expressed as the union of two digraphs, defined in the obvious way. This is equivalent to saying that the underlying graph of D is a connected graph. D is **strongly connected** if, for any two vertices v and w of D, there is a path from v to w. Every strongly connected digraph is connected, but not all connected digraphs are strongly connected; for example, the connected digraph of Fig. 22.1 is not strongly connected since there is no path from v to z.

The distinction between a connected digraph and a strongly connected one becomes clearer if we consider the road map of a city, all of whose streets are one-way. If the road map is connected, then we can drive from any part of the city to any other, ignoring the direction of the one-way streets as we go. If the map is strongly connected, then we can drive from any part of the city to any other, always going the 'right way' down the one-way streets.

Since every one-way system should be strongly connected, it is natural to ask when we can impose a one-way system on a street map in such a way that we can drive from any part of the city to any other. If, for example, the city consists of two parts connected only by a bridge, then we cannot impose such a one-way system on the city, since whatever direction we give to the bridge, one part of the city must be cut off. If, on the other hand, there are no bridges, then we can always impose such a one-way system. This result is stated formally in Theorem 22.1.

For convenience, we define a graph G to be **orientable** if each edge of G can be directed so that the resulting digraph is strongly connected. For example, if G is the graph shown in Fig. 22.5, then G is orientable, since its edges can be directed to give the strongly connected digraph of Fig. 22.6.

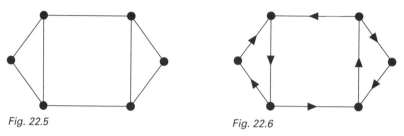

Fig. 22.5   Fig. 22.6

Note that any Eulerian graph is orientable, since we simply follow any Eulerian trail, directing the edges in the direction of the trail as we go. We now give a necessary and sufficient condition (due to H.E. Robbins) for a graph to be orientable.

> THEOREM 22.1. *Let G be a connected graph. Then G is orientable if and only if each edge of G is contained in at least one cycle.*

Proof. The necessity of the condition is clear. To prove the sufficiency, we choose any cycle C and direct its edges cyclically. If each edge of G is contained in C, then the proof is complete. If not, we choose any edge e that is not in C but which is adjacent to an edge of C. By hypothesis, e is contained in some cycle C' whose edges we may direct cyclically, except for those edges that have already been directed – that is, those edges of C' that also lie in C. It is not difficult to see that the resulting digraph is strongly connected; the situation is illustrated in Fig. 22.7, with dashed lines denoting edges of C'. We proceed in this way, at each stage directing at least one new edge, until all edges are directed. Since the digraph remains strongly connected at each stage, the result follows. //

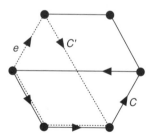

Fig. 22.7

We conclude this section by discussing a 'critical path' problem relating to the scheduling of a series of operations. Suppose that we have a job to perform, such as the building of a house, and that this job can be divided into a number of activities, such as laying the foundations, putting on the roof, doing the wiring, etc. Some of these activities can be performed simultaneously, whereas some may need to be completed before others can be started. Can we find an efficient method for determining which activities should be performed at which times so that the entire job is completed in minimum time?

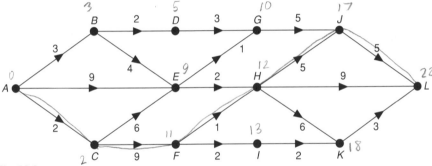

Fig. 22.8

In order to solve this problem, we construct a 'weighted digraph', or **activity network**, in which each arc represents the length of time taken for an activity. Such a network is given in Fig. 22.8. The vertex A represents the beginning of the job, and the vertex L represents its completion. Since the entire job cannot be completed until each path from A to L has been traversed, the problem reduces to that of finding the longest path from A to L. This is accomplished by using a technique known as programme evaluation and review technique (PERT), which is similar to that we used for the shortest path problem in Section 8, except that as we move across the digraph from left to right, we associate with each vertex V a number $l(V)$ indicating the length of the *longest* path from A to V. So for the digraph of Fig. 22.8, we assign:

to vertex A, the number 0;
to vertex B, the number $l(A) + 3$ – that is, 3;
to vertex C, the number $l(A) + 2$ – that is, 2;
to vertex D, the number $l(B) + 2$ – that is, 5;
to vertex E, the number max $\{l(A) + 9, l(B) + 4, l(C) + 6\}$ – that is, 9;
to vertex F, the number $l(C) + 9$ – that is, 11;
to vertex G, the number max $\{l(D) + 3, l(E) + 1\}$ – that is, 10;
to vertex H, the number max $\{l(E) + 2, l(F) + 1\}$ – that is, 12;
to vertex I, the number $l(F) + 2$ – that is, 13;
to vertex J, the number max $\{l(G) + 5, l(H) + 5\}$ – that is, 17;
to vertex K, the number max $\{l(H) + 6, l(I) + 2\}$ – that is, 18;
to vertex L, the number max $\{l(H) + 9, l(J) + 5, l(K) + 3\}$ – that is, 22.

As in the shortest path problem, we keep track of these numbers by writing each one next to the vertex it represents. Note that, unlike the problem that we considered in Section 8, there is no 'zig-zagging', since all arcs are directed from left to right. Thus, the longest path has length 22, and is given in Fig. 22.9. The job cannot therefore be completed until time 22.

This longest path is often called a **critical path**, since any delay in an activity on this path creates a delay in the whole job. In scheduling a job, we therefore need to pay particular attention to the critical paths.

We can also calculate the latest time by which any given operation must be completed if the work is not to be delayed. Working back from L, we see that we must reach K by time $22 - 3 = 19$, J by time $22 - 5 = 17$, H by time min $\{17 - 5, 22 - 9, 19 - 6\} = 12$, and so on.

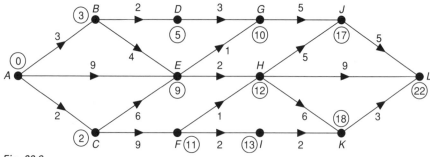

Fig. 22.9

Exercises 22

22.1ˢ   Two of the digraphs in Fig. 22.10 are isomorphic. Which two are they?

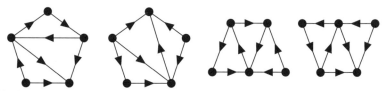

Fig. 22.10

22.2ˢ   Let D be a simple digraph with n vertices and m arcs.
    (i)   Prove that if D is connected, then $n - 1 \le m \le n(n - 1)$.
    (ii) Obtain corresponding bounds for m if D is strongly connected.

22.3ˢ   Write down adjacency matrices for the digraphs in Figs 22.1 and 22.6.

22.4   The **converse** $\tilde{D}$ of a digraph D is obtained by reversing the direction of each arc of D.
    (i)   Give an example of a digraph that is isomorphic to its converse.
    (ii) What is the connection between the adjacency matrices of D and $\tilde{D}$?

22.5   (i)   Without using Theorem 22.1, prove that every Hamiltonian graph is orientable.
    (ii) Show, by finding an orientation for each, that K_n ($n \ge 3$) and $K_{r,s}$ ($r, s \ge 2$) are orientable.
    (iii) Find orientations for the Petersen graph and the graph of the dodecahedron.

22.6ˢ   In the above scheduling problem, calculate the latest times at which we can reach the vertices G, E and B.

22.7   Find the longest path from A to G in the network of Fig. 22.11.

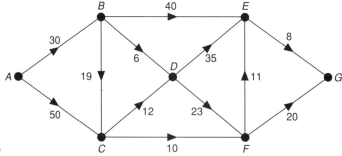

Fig. 22.11

23   Eulerian digraphs and tournaments

In this section we obtain digraph analogues of some results of Sections 6 and 7. In particular, we study Hamiltonian cycles in a type of digraph called a tournament.

A connected digraph D is **Eulerian** if there exists a closed trail containing every arc of D. Such a trail is an **Eulerian trail**. For example, the digraph in Fig. 23.1 is not Eulerian, although its underlying graph is an Eulerian graph.

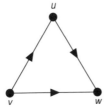

Fig. 23.1

Our first aim is to give a necessary and sufficient condition, analogous to the one in Theorem 6.2, for a connected digraph to be Eulerian. Note that a necessary condition is that the digraph is strongly connected.

We need some preliminary definitions. The **out-degree** of a vertex v of D is the number of arcs of the form vw, and is denoted by outdeg(v). Similarly, the **in-degree** of v is the number of arcs of D of the form wv, and is denoted by indeg(v). Note that the sum of the out-degrees of all the vertices of D is equal to the sum of their in-degrees, since each arc of D contributes exactly 1 to each sum. We call this result the **handshaking dilemma**!

For later convenience, we define a **source** of D to be a vertex with in-degree 0, and a **sink** of D to be a vertex with out-degree 0. Thus, in Fig. 23.1, v is a source and w is a sink. Note that any Eulerian digraph with at least one arc has no sources or sinks. We can now state the basic theorem on Eulerian digraphs.

THEOREM 23.1. *A connected digraph is Eulerian if and only if for each vertex v of D* outdeg(v) = indeg (v).

Proof. The proof is entirely analogous to that of Theorem 6.2 and is left as an exercise. //

We leave it to you to define a semi-Eulerian digraph, and to prove results analogous to Corollaries 6.3 and 6.4.

The corresponding study of Hamiltonian digraphs is, as may be expected, less successful than for Eulerian digraphs. A digraph D is **Hamiltonian** if there is a cycle that includes every vertex of D. A non-Hamiltonian digraph that contains a path passing through every vertex is **semi-Hamiltonian**. Little is known about Hamiltonian digraphs, and several theorems on Hamiltonian graphs do not generalize easily, if at all, to digraphs.

It is natural to ask whether there is a generalization to digraphs of Dirac's theorem (Corollary 7.2). One such generalization is due to Ghouila-Houri; its proof is considerably more difficult than that of Dirac's theorem, and can be found in Bondy and Murty [7].

THEOREM 23.2. *Let D be a strongly connected digraph with n vertices. If* outdeg(v) $\geq$ n/2 *and* indeg(v) $\geq$ n/2 *for each vertex v, then D is Hamiltonian.*

It seems that such results will not come easily, and so we consider instead which types of digraph are Hamiltonian. In this respect, the tournaments are particularly important, the results in this case taking a very simple form.

A **tournament** is a digraph in which any two vertices are joined by exactly one arc (see Fig. 23.2). Such a digraph can be used to record the result of a tennis tournament, or any other game in which draws are not allowed. In Fig. 23.2, for example, team z beats team w, but is beaten by team v, and so on.

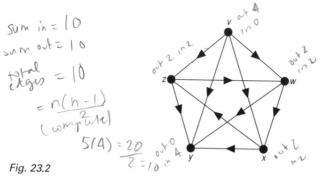

Fig. 23.2

Because tournaments may have sources or sinks, they are not in general Hamiltonian. However, the following theorem, due to L. Rédei and P. Camion, shows that every tournament is 'nearly Hamiltonian'.

THEOREM 23.3. *(i) Every non-Hamiltonian tournament is semi-Hamiltonian; (ii) every strongly connected tournament is Hamiltonian.*

Proof. (i) The statement is clearly true if the tournament has fewer than four vertices. We prove the result by induction on the number of vertices, and assume that every non-Hamiltonian tournament on n vertices is semi-Hamiltonian.

Let T be a non-Hamiltonian tournament on $n + 1$ vertices, and let T' be the tournament on n vertices obtained by removing from T a vertex v and its incident arcs. By the induction hypothesis, T' has a semi-Hamiltonian path $v_1 \to v_2 \to \cdots \to v_n$. There are now three cases to consider:

(1) if vv_1 is an arc in T, then the required path is

$$v \to v_1 \to v_2 \cdots \to v_n.$$

(2) if vv_1 is not an arc in T, which means that v_1v is, and if there exists an i such that vv_i is an arc in T, then choosing i to be the first such, the required path is (see Fig. 23.3)

$$v_1 \to v_2 \to \cdots \to v_{i-1} \to v \to v_i \to \cdots \to v_n.$$

(3) if there is no arc in T of the form vv_i, then the required path is

$$v_1 \to v_2 \to \cdots \to v_n \to v.$$

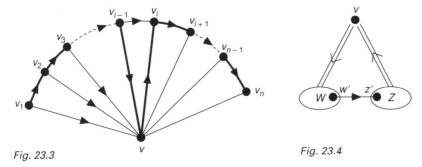

Fig. 23.3   Fig. 23.4

(ii) We prove the stronger result that a strongly connected tournament T on n vertices contains cycles of length $3, 4, \ldots, n$. To show that T contains a cycle of length 3, let v be any vertex of T, and let W be the set of all vertices w such that vw is an arc in T, and Z be the set of all vertices z such that zv is an arc. Since T is strongly connected, W and Z must both be non-empty, and there must be an arc in T of the form $w'z'$, where w' is in W and z' is in Z (see Fig. 23.4). The required cycle of length 3 is then $v \to w' \to z' \to v$.

It remains only to show that, if there is a cycle of length k, where $k \leq n$, then there is one of length $k + 1$. Let $v_1 \to \cdots \to v_k \to v_1$ be such a cycle. Suppose first that there exists a vertex v not contained in this cycle, such that there exist arcs in T of the form vv_i and of the form v_jv. Then there must be a vertex v_i such that both $v_{i-1}v$ and vv_i are arcs in T. The required cycle is then (see Fig. 23.5)

$$v_1 \to v_2 \to \cdots \to v_{i-1} \to v \to v_i \to \cdots \to v_k \to v_1.$$

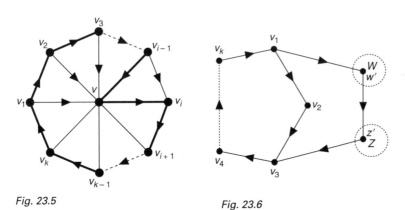

Fig. 23.5

Fig. 23.6

If no vertex exists with the above-mentioned property, then the set of vertices not contained in the cycle may be divided into two disjoint sets W and Z, where W is the set of vertices w such that v_iw is an arc for each i, and Z is the set of vertices z such that zv_i is an arc for each i. Since T is strongly connected, W and Z must both be non-empty, and there must be an arc in T of the form $w'z'$, where w' is in W and z' is in Z. The required cycle is then (see Fig. 23.6)

$$v_1 \to w' \to z' \to v_3 \to \cdots \to v_k \to v_1. \; /\!/$$

Exercises 23

23.1ˢ   Verify the handshaking dilemma for the tournaments of Figs. 23.2 and 23.7.

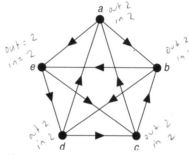

Fig. 23.7

23.2ˢ   In the tournament of Fig. 23.7, find
    (i)   cycles of length 3, 4 and 5;
    (ii) an Eulerian trail;
    (iii) a Hamiltonian cycle.

23.3ˢ   Prove that a tournament cannot have more than one source or more than one sink.

23.4   Let T be a tournament on n vertices. If Σ denotes a summation over all the vertices of T, prove that
    (i) $\Sigma \, \mathrm{outdeg}(v) = \Sigma \, \mathrm{indeg}(v)$;
    (ii) $\Sigma \, \mathrm{outdeg}(v)^2 = \Sigma \, \mathrm{indeg}(v)^2$.

23.5   Let D be the digraph whose vertices are the pairs of integers
       11, 12, 13, 21, 22, 23, 31, 32, 33,
and whose arcs join ij to kl if and only if $j = k$. Find an Eulerian trail in D and use it to obtain a circular arrangement of nine 1s, nine 2s and nine 3s in which each of the 27 possible triples (111, 233, etc.) occurs exactly once. (Problems of this kind arise in communication theory.)

23.6   A tournament T is **irreducible** if it is impossible to split the set of vertices of T into two disjoint sets V_1 and V_2 so that each arc joining a vertex of V_1 and a vertex of V_2 is directed from V_1 to V_2.
    (i)   Give an example of an irreducible tournament.
    (ii) Prove that a tournament is irreducible if and only if it is strongly connected.

23.7   A tournament is **transitive** if the existence of arcs uv and vw implies the existence of the arc uw.
    (i)   Give an example of a transitive tournament.
    (ii) Show that in a transitive tournament the teams can be ranked so that each team beats all the teams which follow it in the ranking.
    (iii) Deduce that a transitive tournament with at least two vertices cannot be strongly connected.

23.8*   The **score** of a vertex of a tournament is its out-degree, and the **score sequence** of a tournament is the sequence formed by arranging the scores of its vertices in nondecreasing order; for example, the score-sequence of the tournament in Fig. 23.2 is $(0, 2, 2, 2, 4)$. Show that if $(s_1, \ldots, s_n)$ is the score-sequence of a tournament T, then
    (i) $s_1 + \cdots + s_n = n(n - 1)/2$;
    (ii) for each positive integer $k < n$, $s_1 + \cdots + s_k \geq k(k - 1)/2$, with strict inequality for all k if and only if T is strongly connected;
    (iii) T is transitive if and only if $s_k = k - 1$ for each k.

24 Markov chains

As you have already seen, digraphs turn up in a variety of practical situations. In this section we describe a simple application of digraphs to the study of finite Markov chains. Another application – the study of flows in networks – is discussed in the next chapter. For further applications, see Deo [13] or Wilson and Beineke [21].

The study of Markov chains has arisen in a wide variety of areas, ranging from genetics and statistics to computing and sociology. For ease of presentation, we consider a rather trivial problem, that of the drunkard standing directly between his two favourite pubs, 'The Markov Chain' and 'The Source and Sink' (see Fig. 24.1).

Fig. 24.1

Every minute he either staggers ten metres towards the first pub (with probability $\frac{1}{2}$)) or towards the second pub (with probability $\frac{1}{3}$) or he stays where he is (with probability $\frac{1}{6}$) – such a procedure is called a one-dimensional **random walk**. We assume that the two pubs are 'absorbing', in the sense that if he arrives at either of them he stays there. Given the distance between the two pubs and his initial position, we can ask which pub he is more likely to reach first, and how long he is likely to take getting there.

Let us suppose that the two pubs are 50 metres apart, and that our friend is initially 20 metres from 'The Source and Sink'. If we denote the places at which he can stop by $E_1, \ldots, E_6$, where E_1 and E_6 are the two pubs, then his initial position E_4 can be described by the vector $\mathbf{x} = (0, 0, 0, 1, 0, 0)$, in which the ith component is the probability that he is initially at E_i. Furthermore, the probabilities of his position after one minute are given by the vector $(0, 0, \frac{1}{2}, \frac{1}{6}, \frac{1}{3}, 0)$, and after two minutes by $(0, \frac{1}{4}, \frac{1}{6}, \frac{13}{36}, \frac{1}{9}, \frac{1}{9})$. It is awkward to calculate directly the probability of his being at a given place after k minutes, and a more convenient way of doing this is to introduce the transition matrix.

Let p_{ij} be the probability that he moves from E_i to E_j in one minute; for example, $p_{23} = \frac{1}{3}$ and $p_{24} = 0$. These probabilities p_{ij} are called the **transition probabilities**, and the 6×6 matrix $\mathbf{P} = (p_{ij})$ is the **transition matrix** (see Fig. 24.2).

$$\begin{pmatrix} 1 & 0 & 0 & 0 & 0 & 0 \\ \frac{1}{2} & \frac{1}{6} & \frac{1}{3} & 0 & 0 & 0 \\ 0 & \frac{1}{2} & \frac{1}{6} & \frac{1}{3} & 0 & 0 \\ 0 & 0 & \frac{1}{2} & \frac{1}{6} & \frac{1}{3} & 0 \\ 0 & 0 & 0 & \frac{1}{2} & \frac{1}{6} & \frac{1}{3} \\ 0 & 0 & 0 & 0 & 0 & 1 \end{pmatrix}$$

Fig. 24.2

Note that each entry of **P** is non-negative and that the sum of the entries in each row is 1. If **x** is the initial row vector defined above, then the probabilities of his position after one minute are given by the row vector **xP**, and after k minutes by the vector **xP**^{k}. In other words, the ith component of **xP**^{k} represents the probability that he is at E_i after k minutes have elapsed.

In general, we define a **probability vector** to be a row vector whose entries are all non-negative and have sum 1, and a **transition matrix** to be a square matrix, each of whose rows is a probability vector. We then define a finite **Markov chain** (or simply, a **chain**) to consist of an $n \times n$ transition matrix **P** and a $1 \times n$ row vector **x**. The positions E_i are the **states** of the chain, and our aim is to describe ways of classifying them.

We are mainly concerned with whether we can get from a given state to another state, and if so, how long it takes. For example, in the above problem, the drunkard can get from E_4 to E_1 in three minutes and from E_4 to E_6 in two minutes, but cannot get from E_1 to E_4, because the pub E_1 is 'absorbing'. Thus, our main concern is not with the actual probabilities p_{ij}, but with when they are non-zero. To decide this, we represent the situation by a digraph whose vertices correspond to the states and whose arcs tell us whether we can go from one state to another in one minute. Thus, if each state E_i is represented by a vertex v_i, then we obtain the required digraph by drawing an arc from v_i to v_j if and only if $p_{ij} \neq 0$. Alternatively, we can define the digraph in terms of its adjacency matrix by replacing each non-zero entry of the matrix **P** by 1. We refer to this digraph as the **associated digraph** of the Markov chain. The associated digraph of the above problem is shown in Fig. 24.3.

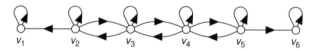

Fig. 24.3

As a further example, if we are given a Markov chain whose transition matrix is the matrix of Fig. 24.4, then its associated adjacency matrix and digraph are as shown in Fig. 24.5.

$$\begin{pmatrix} 0 & \frac{1}{4} & \frac{1}{2} & 0 & 0 & \frac{1}{4} \\ 0 & 1 & 0 & 0 & 0 & 0 \\ \frac{1}{2} & \frac{1}{3} & 0 & \frac{1}{12} & 0 & \frac{1}{12} \\ 0 & 0 & 0 & 0 & 1 & 0 \\ 0 & 0 & 0 & 0 & 0 & 1 \\ 0 & 0 & 0 & 1 & 0 & 0 \end{pmatrix} \qquad \begin{pmatrix} 0 & 1 & 1 & 0 & 0 & 1 \\ 0 & 1 & 0 & 0 & 0 & 0 \\ 1 & 1 & 0 & 1 & 0 & 1 \\ 0 & 0 & 0 & 0 & 1 & 0 \\ 0 & 0 & 0 & 0 & 0 & 1 \\ 0 & 0 & 0 & 1 & 0 & 0 \end{pmatrix}$$

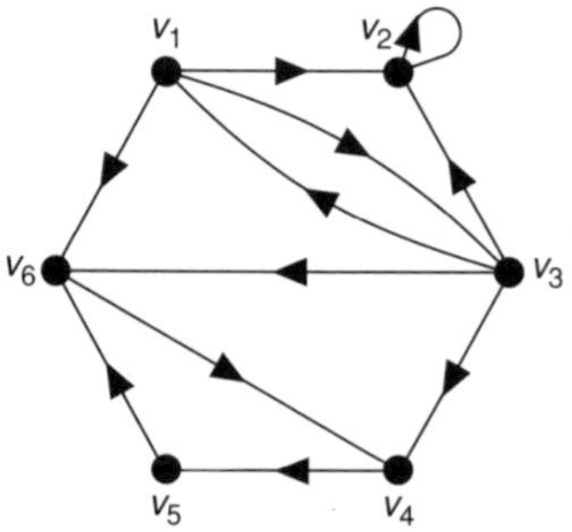

Fig. 24.4   *Fig. 24.5*

Note that we can get from a state E_i to a state E_j in a Markov chain if and only if there is a path from v_i to v_j in the associated digraph, and the least possible time to do so is the length of the shortest such path. A Markov chain in which we can get from any state to any other is called an **irreducible chain**. Clearly a Markov chain is irreducible if and only if its associated digraph is strongly connected. Note that neither of the Markov chains described above is irreducible; for example, in the second chain there is no path from v_2 to any other vertex.

In investigating these matters further, we distinguish between those states to which we keep on returning however long we continue, and those that we visit a few times and then never return to. More formally, if on starting at E_i the probability of returning to E_i at some later stage is 1, then E_i is a **persistent** state; otherwise E_i is **transient**. For example, in the drunkard problem, E_1 and E_6 are persistent and the other states are transient. In complicated examples, the calculation of the relevant probabilities can be tricky, and it is often easier to classify the states by analysing the associated digraph of the chain. It is easy to see that a state E_i is persistent if and only if the existence of a path from v_i to v_j in the associated digraph implies the existence of a path from $v_j \to v_i$. In Fig. 24.5 there is a path from v_1 to v_4 but no path from v_4 to v_1. Thus E_1 is transient, as is E_3, whereas E_2, E_4, E_5 and E_6 are persistent. A state, such as E_2, from which we can get to no other state is an **absorbing** state.

An alternative way of classifying states is in terms of their periodicity. A state E_i of a Markov chain is **periodic of period t** $(t \neq 1)$ if it is possible to return to E_i only after a period of time that is a multiple of t; if no such t exists, then E_i is **aperiodic**. Note that every state E_i for which $p_{ii} \neq 0$ is aperiodic; for example, every absorbing state is aperiodic. In the drunkard problem, the absorbing states E_1 and E_6 are not the only aperiodic states – in fact, every state is aperiodic. In the second example, the absorbing state E_2 is the only aperiodic state, since E_1 and E_3 are periodic of period 2 and E_4, E_5 and E_6 are periodic of period 3. In digraph terms, a state E_i is periodic of period t if and only if in the associated digraph the length of each closed trail containing v_i is a multiple of t. Finally, a state is **ergodic** if it is both persistent and aperiodic, and if every state is ergodic then the chain is an **ergodic chain**. For many purposes, ergodic chains are the most important and desirable chains to deal with. An example of such a chain is given in Exercise 24.2.

Exercises 24

24.1^s   (i)   Suppose that, in the drunkard problem, the right-hand pub ejects him as soon as he gets there. Write down the resulting transition matrix and its associated digraph, and re-classify the states.

(ii)   How would your answers to part (i) be changed if *both* pubs eject him?

24.2^s   A game is played with a die by five people around a circular table. If the player with the die throws an odd number, he passes the die to the player on his left; if he throws a 2 or 4, he passes it to the player on his right; if he throws a 6, he keeps the die and throws again.

(i)   Write down the corresponding transition matrix and its associated digraph.

(ii)   Show that each state is persistent and aperiodic, and deduce that the corresponding Markov chain is ergodic.

24.3   (i)   Prove that, if $\mathbf{P}$ and $\mathbf{Q}$ are transition matrices, then so is $\mathbf{PQ}$.

(ii)   What is the connection between the associated digraphs of $\mathbf{P}$ and $\mathbf{Q}$ and that of $\mathbf{PQ}$?

24.4*   (i)   Prove that every finite Markov chain has at least one persistent state.

(ii)   Deduce that, if a finite Markov chain is irreducible, then each state is persistent.

(iii)   Show how infinite Markov chains can be defined, and construct one in which each state is transient.

Chapter 8

Matching, marriage and Menger's theorem

*They drew all manner of things—
everything that begins with an M—.*
Lewis Carroll

The results of this chapter are more combinatorial than those of the preceding chapters, although we shall see that they are closely related to graph theory. We begin with a discussion of Hall's 'marriage' theorem in several different contexts, including its applications to the construction of latin squares and timetabling problems. In Section 28 we present a theorem of Menger on the number of disjoint paths connecting a given pair of vertices in a graph or digraph. In Section 29 we give an alternative formulation of Menger's theorem, known as the *max-flow min-cut theorem,* which is of fundamental importance in connection with network flow problems.

25   Hall's 'marriage' theorem

The marriage theorem, proved in 1935 by Philip Hall, answers the following question, known as the **marriage problem**: *if there is a finite set of girls, each of whom knows several boys, under what conditions can all the girls marry the boys in such a way that each girl marries a boy she knows?* For example, if there are four girls $\{g_1, g_2, g_3, g_4\}$ and five boys $\{b_1, b_2, b_3, b_4, b_5\}$, and the friendships are as shown in Fig. 25.1, then a possible solution is for g_1 to marry b_4, g_2 to marry b_1, g_3 to marry b_3, and g_4 to marry b_2.

girl	boys known by girl		
g_1	b_1	b_4	b_5
g_2	b_1		
g_3	b_2	b_3	b_4
g_4	b_2	b_4	

Fig. 25.1

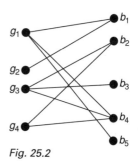

Fig. 25.2

This problem can be represented graphically by taking G to be the bipartite graph in which the vertex set is divided into two disjoint sets V_1 and V_2, corresponding to the girls and boys, and where each edge joins a girl to a boy she knows. Fig. 25.2 shows the graph G corresponding to the situation of Fig. 25.1.

A **complete matching** from V_1 to V_2 in a bipartite graph $G(V_1, V_2)$ is a one–one correspondence between the vertices in V_1 and a subset of the vertices in V_2, such that corresponding vertices are joined. The marriage problem can then be expressed in graph-theoretic terms in the form: *if $G = G(V_1, V_2)$ is a bipartite graph, when does there exist a complete matching from V_1 to V_2 in G?*

Returning to 'matrimonial terminology', we note that, for the solution of the marriage problem, every k girls must know collectively at least k boys, for all integers k satisfying $1 \le k \le m$, where m denotes the total number of girls. We refer to this condition as the **marriage condition**. It is a necessary condition because, if it were not true for a given set of k girls, then we could not marry the girls in that set, let alone the others.

Surprisingly, the marriage condition also turns out to be sufficient. This is the content of Hall's 'marriage' theorem. Because of its importance, we give three proofs; the first is due to P. Halmos and H.E. Vaughan.

> **THEOREM 25.1** (Hall, 1935). *A necessary and sufficient condition for a solution of the marriage problem is that each set of k girls collectively knows at least k boys, for $1 \le k \le m$.*

Remark. Although this theorem is couched in the frivolous terms of the marriage problem, it also applies to more serious problems. For example, it gives a necessary and sufficient condition for the solution of a job assignment problem in which applicants must be assigned to jobs for which they are variously qualified. An example of such a problem is given in Exercise 25.2.

Proof. As noted above, the condition is necessary. To prove that it is sufficient, we use induction on m, and assume that the theorem is true if the number of girls is *less than* m. Note that the theorem is true if $m = 1$.

Suppose now that there are m girls. There are two cases to consider.

(i) If every k girls (where $k < m$) collectively know at least $k + 1$ boys, so that the condition is always true 'with one boy to spare', then we take any girl and marry her to any boy she knows. The original condition then remains true for the other $m - 1$ girls, who can be married by induction, completing the proof in this case.

(ii) If now there is a set of k girls ($k < m$) who collectively know *exactly* k boys, then these k girls can be married by induction to the k boys, leaving $m - k$ girls still to be married. But any collection of h of these $m - k$ girls, for $h \le m - k$, must know at least h of the remaining boys, since otherwise these h girls, together with the above collection of k girls, would collectively know fewer than $h + k$ boys, contrary to our assumption. It follows that the original condition applies to the $m - k$ girls. They can therefore be married by induction in such a way that everyone is happy and the proof is complete. //

We can also state Hall's theorem in the language of complete matchings in a bipartite graph. Recall that the number of elements in a set S is denoted by $|S|$.

> **COROLLARY 25.2.** *Let $G = G(V_1, V_2)$ be a bipartite graph, and for each subset A of V_1, let $\varphi(A)$ be the set of vertices of V_2 that are adjacent to at least one vertex of A. Then a complete matching from V_1 to V_2 exists if and only if $|A| \leq |\varphi(A)|$, for each subset A of V_1.*

Proof. The proof of this corollary is a translation into graph terminology of the above proof. //

Exercises 25

25.1[s] Suppose that three boys a, b, c know four girls w, x, y, z as in Fig 25.3:

boy	girls known by boy		
a	w	y	z
b	x	z	
c	x	y	

Fig. 25.3

   (i)   Draw the bipartite graph corresponding to this table of relationships.
   (ii)   Find five different solutions of the corresponding marriage problem.
   (iii)   Check the marriage condition for this problem.

25.2 A building contractor advertises for a bricklayer, a carpenter, a plumber and a tool-maker, and receives five applicants – one for the job of bricklayer, one for carpenter, one for bricklayer and plumber, and two for plumber and toolmaker.
   (i)   Draw the corresponding bipartite graph.
   (ii)   Check whether the marriage condition holds for this problem.
Can all of the jobs be filled by qualified people?

25.3[s] Explain why the graph in Fig. 25.4 has no complete matching from V_1 to V_2. When does the marriage condition fail?

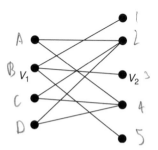

Fig. 25.4

25.4 (The 'harem problem') Let B be a set of boys, and suppose that each boy in B wishes to marry more than one of his girl friends. Find a necessary and sufficient condition for the harem problem to have a solution. (Hint: replace each boy by several identical copies of himself, and then use Hall's theorem.)

25.5   Prove that, if $G = G(V_1, V_2)$ is a bipartite graph in which the degree of each vertex in V_1 is not less than the degree of each vertex in V_2, then G has a complete matching.

25.6*   (i)   Use the marriage condition to show that if each girl has r (≥ 1) boy friends and each boy has r girl friends, then the marriage problem has a solution.

(ii)   Use the result of part (i) to prove that, if G is a bipartite graph which is regular of degree r, then G has a complete matching. Deduce that the chromatic index of G is r. (This is a special case of Theorem 20.4.)

25.7*   Suppose that the marriage condition is satisfied, and that each of the m girls knows at least t boys. Show, by induction on m, that the marriages can be arranged in at least $t!$ ways if $t \leq m$, and in at least $t!/(t-m)!$ ways if $t > m$.

26   Transversal theory

This section is devoted to an alternative proof of Hall's theorem, given in the language of transversal theory. We leave the translation of this proof into matching or marriage terminology as an exercise.

Recall from our example in the previous section (see Fig. 25.1) that the sets of boys known by the four girls were $\{b_1, b_4, b_5\}$, $\{b_1\}$, $\{b_2, b_3, b_4\}$ and $\{b_2, b_4\}$, and that a solution of the marriage problem was obtained by finding four distinct bs, one from each of these sets of boys (see Fig. 26.1).

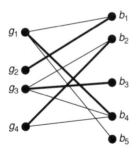

Fig. 26.1

In general, if E is a non-empty finite set, and if $\mathcal{F} = (S_1, \ldots, S_m)$ is a family of (not necessarily distinct) non-empty subsets of E, then a **transversal** of $\mathcal{F}$ is a set of m distinct elements of E, one chosen from each set S_i.

Now suppose that $E = \{1, 2, 3, 4, 5, 6\}$, and let

$$S_1 = S_2 = \{1, 2\}, S_3 = S_4 = \{2, 3\}, S_5 = \{1, 4, 5, 6\}.$$

Then it is impossible to find five distinct elements of E, one from each subset S_i; in other words, the family $\mathcal{F} = (S_1, \ldots, S_5)$ has no transversal. However, the subfamily $\mathcal{F}' = (S_1, S_2, S_3, S_5)$ has a transversal – for example, $\{1, 2, 3, 4\}$. We call a transversal of a subfamily of $\mathcal{F}$ a **partial transversal** of $\mathcal{F}$. In this example, $\mathcal{F}$ has several partial transversals, such as $\{1, 2, 3, 6\}$, $\{2, 3, 6\}$, $\{1, 5\}$, and $\varnothing$. Note that any subset of a partial transversal is a partial transversal.

It is natural to ask under what conditions a given family of subsets of a set has a transversal. The connection between this problem and the marriage problem is easily seen by taking E to be the set of boys, and S_i to be the set of boys known by girl g_i, for

$1 \leq i \leq m$. A transversal in this case is then simply a set of m boys, one corresponding to, and known by, each girl. It follows that Theorem 25.1 gives a necessary and sufficient condition for a given family of sets to have a transversal.

We restate Hall's theorem in this form, and give an alternative proof due to R. Rado. The beauty of this proof lies in the fact that essentially only one step is involved, in contrast to the Halmos–Vaughan proof which involves two separate cases. It is, however, more awkward to express this proof in the intuitive and appealing language of matrimony!

THEOREM 26.1. *Let E be a non-empty finite set, and let $\mathcal{F} = (S_1, \ldots, S_m)$ be a family of non-empty subsets of E. Then $\mathcal{F}$ has a transversal if and only if the union of any k of the subsets S_i contains at least k elements $(1 \leq k \leq m)$.*

Proof. The necessity of the condition is clear. To prove the sufficiency, we show that if one of the subsets (S_1, say) contains more than one element, then we can remove an element from S_1 without altering the condition. By repeating this procedure, we eventually reduce the problem to the case in which each subset contains only one element, and the proof is then trivial.

It remains only to show the validity of this 'reduction procedure'. So, suppose that S_1 contains elements x and y, the removal of either of which invalidates the condition. Then there are subsets A and B of $\{2, 3, \ldots, m\}$ with the property that $|P| \leq |A|$ and $|Q| \leq |B|$, where

$$P = \bigcup_{j \in A} S_j \cup (S_1 - \{x\}) \quad \text{and} \quad Q = \bigcup_{j \in B} S_j \cup (S_1 - \{y\}).$$

Then

$$|P \cup Q| = |\bigcup_{j \in A \cup B} S_j \cup S_1| \quad \text{and} \quad |P \cup Q| \geq |\bigcup_{j \in A \cap B} S_j|$$

The required contradiction now follows, since

$$|A| + |B| \geq |P| + |Q|$$
$$= |P \cup Q| + |P \cap Q|$$
$$\geq |\bigcup_{j \in A \cup B} S_j \cup S_1| + |\bigcup_{j \in A \cap B} S_j|$$
$$\geq (|A \cup B| + 1) + |A \cap B| \quad \text{(by Hall's condition)}$$
$$= |A| + |B| + 1. \;//$$

Before proceeding to some applications of Hall's theorem, we state two corollaries that will be needed in Section 33. In marriage terminology, the first of these corollaries gives us a condition under which at least t girls can marry boys known to them.

COROLLARY 26.2. *If E and $\mathcal{F}$ are as before, then $\mathcal{F}$ has a partial transversal of size t if and only if the union of any k of the subsets S_i contains at least $k + t - m$ elements.*

Sketch of proof. The result follows on applying Theorem 26.1 to the family $\mathcal{F}' = (S_1 \cup D, \ldots, S_m \cup D)$, where D is any set disjoint from E and containing $m - t$ elements. Note that $\mathcal{F}$ has a partial transversal of size t if and only if $\mathcal{F}'$ has a transversal. //

COROLLARY 26.3. *If E and $\mathcal{F}$ are as before, and if X is any subset of E, then $\mathcal{F}X$ contains a partial transversal of $\mathcal{F}$ of size t if and only if, for each subset A of $\{1, \ldots, m\}$,* $\qquad$ *Fx is family*

$$\left| \left(\bigcup_{j \in A} S_j \right) \cap X \right| \geq |A| + t - m.$$

Sketch of proof. The result follows on applying the previous corollary to the family $\mathcal{F}_X = (S_1 \cap X, \ldots, S_m \cap X)$. //

Exercises 26

26.1[s] Decide which of the following families of subsets of $E = \{1, 2, 3, 4, 5\}$ have transversals, list all the transversals of those that have transversals, and list all the partial transversals of those that have no transversal:
  (i) $(\{1\}, \{2, 3\}, \{1, 2\}, \{1, 3\}, \{1, 4, 5\})$;
  (ii) $(\{1, 2\}, \{2, 3\}, \{4, 5\}, \{4, 5\})$; $\{1, 2, 4, 5\}, \{1, 3, 4, 5\} \{1 2 5 4\} \{1 3 5 4\}$
  (iii) $(\{1, 3\}, \{2, 3\}, \{1, 2\}, \{3\})$; $\qquad \{2 3 5 4\} \{2 3 4 5\}$
  (iv) $(\{1, 3, 4\}, \{1, 4, 5\}, \{2, 3, 5\}, \{2, 4, 5\})$.

26.2 Repeat Exercise 26.1 for the set $\{G, R, A, P, H, S\}$:
  (i) $(\{R\}, \{R, G\} \{A, P\}, \{A, H\}, \{R, A\})$;
  (ii) $(\{R\}, \{R, G\}, \{A, G\}, \{A, R\})$;
  (iii) $(\{G, R\}, \{R, P, H\}, \{G, S\}, \{R, H\})$;
  (iv) $(\{R, P\}, \{R, P\}, \{R, G\}, \{R\})$.

26.3[s] Let E be the set of letters in the word *MATROIDS*. Show that the family *(STAR, ROAD, MOAT, RIOT, RIDS, DAMS, MIST)* of subsets of E has exactly eight transversals.

26.4[s] Let E be the set $\{1, 2, \ldots, 50\}$. How many distinct transversals has the family $(\{1, 2\}, \{2, 3\}, \{3, 4\}, \ldots, \{50, 1\})$?

26.5 Verify the statements of Corollaries 26.2 and 26.3 when $E = \{a, b, c, d, e\}$ and $\mathcal{F} = (\{a, c, e\}, \{b, d\}, \{b, d\}, \{b, d\})$, and $X = \{a, b, c\}$.

26.6[s] Let $E = \{\blacklozenge, \blacktriangledown, \spadesuit, \clubsuit, \bigstar\}$ and $\mathcal{F} = (\{\blacklozenge, \blacktriangledown, \spadesuit\}, \{\spadesuit, \clubsuit\}, \{\clubsuit\}, \{\clubsuit\}, \{\blacklozenge, \blacktriangledown, \bigstar\})$.
  (i) List all the subfamilies of $\mathcal{F}$ for which the marriage condition is not satisfied.
  (ii) Verify the statement of Corollary 26.2.

26.7 Rewrite
  (i) the statements of Corollaries 26.2 and 26.3 in marriage terminology;
  (ii) the Halmos–Vaughan proof of Hall's theorem in the language of transversal theory.

26.8* Let E and $\mathcal{F}$ have their usual meanings, let T_1 and T_2 be transversals of $\mathcal{F}$, and let x be an element of T_1. Show that there exists an element y of T_2 such that $(T_1 - \{x\}) \cup \{y\}$ (the set obtained from T_1 on replacing x by y) is also a transversal of $\mathcal{F}$. Compare this result with Exercise 9.11.
(This result will be needed in Chapter 9.)

26.9* The **rank** $r(A)$ of a subset A of E is the number of elements in the largest partial transversal of $\mathcal{F}$ contained in A. Show that
  (i) $0 \le r(A) \le |A|$;
  (ii) if $A \subseteq B \subseteq E$, then $r(A) \subseteq r(B)$;
  (iii) if $A, B \subseteq E$, then $r(A \cup B) + r(A \cap B) \le r(A) + r(B)$.
Compare these results with Exercise 9.12.
(This result will also be needed in Chapter 9.)

26.10* Let $\mathcal{F}$ be a family consisting of m non-empty subsets of E, and let A be a subset of E. By applying Hall's theorem to the family consisting of $\mathcal{F}$ together with $|E| - m$ copies of $E - A$, prove that there exists a transversal of $\mathcal{F}$ containing A if and only if
  (i) $\mathcal{F}$ has a transversal;
  (ii) A is a partial transversal of $\mathcal{F}$.
(A simpler proof, using matroids, is given in Section 33.)

26.11* Let E be a countable set, and let $\mathcal{F} = (S_1, S_2, \ldots)$ be a countable family of non-empty *finite* subsets of E.
  (i)   Defining a transversal of $\mathcal{F}$ in the natural way, show, by König's lemma (Theorem 16.3), that $\mathcal{F}$ has a transversal if and only if the union of any k subsets S_i contains at least k elements, for all finite k.
  (ii) By considering $E = \{1, 2, 3, \ldots\}$, $S_1 = E$, $S_2 = \{1\}$, $S_3 = \{2\}$, $S_4 = \{3\}, \ldots$, show that the result of part (i) is false if not all of the S_i are finite.

27   Applications of Hall's theorem

In this section we apply Hall's theorem to problems concerning the construction of Latin squares, the elements of a (0, 1)-matrix, and the existence of a common transversal of two families of subsets of a given set. The last of these applications is of relevance in timetabling problems.

Latin squares
An $m \times n$ **latin rectangle** is an $m \times n$ matrix $\mathbf{M} = (m_{ij})$ whose entries are integers satisfying:
  (i) $1 \le m_{ij} \le n$;
  (ii) no two entries in any row or in any column are equal.

Note that (i) and (ii) imply that $m \le n$. If $m = n$, then the latin rectangle is a **latin square**. For example, Figs 27.1 and 27.2 show a 3×5 latin rectangle and a 5×5 latin square.

$$\begin{pmatrix} 1 & 2 & 3 & 4 & 5 \\ 2 & 4 & 1 & 5 & 3 \\ 2 & 5 & 2 & 1 & 4 \end{pmatrix}$$

Fig. 27.1

$$\begin{pmatrix} 1 & 2 & 3 & 4 & 5 \\ 2 & 4 & 1 & 5 & 3 \\ 3 & 5 & 2 & 1 & 4 \\ 4 & 3 & 5 & 2 & 1 \\ 5 & 1 & 4 & 3 & 2 \end{pmatrix}$$

Fig. 27.2

Given an $m \times n$ latin rectangle with $m < n$, when can we adjoin $n - m$ new rows so that a latin square is produced? Surprisingly, the answer is 'always'!

THEOREM 27.1. *Let* **M** *be an* $m \times n$ *latin rectangle with* $m < n$. *Then* **M** *can be extended to a latin square by the addition of* $n - m$ *new rows.*

Proof. We prove that **M** can be extended to an $(m + 1) \times n$ latin rectangle. By repeating the procedure involved, we eventually obtain a latin square.

Let $E = \{1, 2, \ldots, n\}$ and $\mathcal{F} = (S_1, \ldots, S_n)$, where S_i is the set consisting of those elements of E that do *not* occur in the ith column of **M**. If we can prove that $\mathcal{F}$ has a transversal, then the proof is complete, since the elements in this transversal form the additional row. By Hall's theorem, it is sufficient to show that the union of any k of the S_i contains at least k distinct elements. But this is obvious, since such a union contains $(n - m)k$ elements altogether, including repetitions, and if there were fewer than k distinct elements, then at least one of them would have to appear more than $n - m$ times. Since each element occurs exactly $n - m$ times, we have the required contradiction. //

(0, 1)-matrices

An alternative way of studying transversals of a family $\mathcal{F} = (S_1, \ldots, S_m)$ of non-empty subsets of a set $E = \{e_1, \ldots, e_n\}$ is to study the **incidence matrix** of the family – the $m \times n$ matrix $\mathbf{A} = (a_{ij})$ in which $a_{ij} = 1$ if $e_j \in S_i$, and $a_{ij} = 0$ otherwise. We call such a matrix, in which each entry is 0 or 1, a **(0, 1)-matrix**. If the **term rank** of **A** is the largest number of 1s of **A**, no two of which lie in the same row or column, then $\mathcal{F}$ has a transversal if and only if the term rank of **A** is m. Moreover, the term rank of **A** is precisely the number of elements in a partial transversal of largest possible size. As a second application of Hall's theorem, we prove a famous result on (0, 1)-matrices known as the **König–Egerváry theorem**.

THEOREM 27.2 (König–Egerváry, 1931). *The term rank of a* (0, 1)*-matrix* **A** *is equal to the minimum number* μ *of rows and columns that together contain all the* 1*s of* **A**.

Remark. As an illustration of the theorem, consider the matrix of Fig. 27.3 which is the incidence matrix of the family $\mathcal{F} = (S_1, S_2, S_3, S_4, S_5)$ of subsets of $E = \{1, 2, 3, 4, 5, 6\}$, where $S_1 = S_2 = \{1, 2\}$, $S_3 = S_4 = \{2, 3\}$, $S_5 = \{1, 4, 5, 6\}$. Clearly the term rank and μ are both 4.

$$
\begin{array}{c}
& \begin{array}{cccccc} e_1 & e_2 & e_3 & e_4 & e_5 & e_6 \end{array} \\
\begin{array}{c} S_1 \\ S_2 \\ S_3 \\ S_4 \\ S_5 \end{array}
& \left(\begin{array}{cccccc}
\textcircled{1} & 1 & 0 & 0 & 0 & 0 \\
1 & \textcircled{1} & 0 & 0 & 0 & 0 \\
0 & 1 & 1 & 0 & 0 & 0 \\
0 & 1 & \textcircled{1} & 0 & 0 & 0 \\
1 & 0 & 0 & \textcircled{1} & 1 & 1
\end{array}\right)
\end{array}
$$

Fig. 27.3

Proof. It is clear that the term rank cannot exceed μ. To prove equality, we can suppose that all the 1s of $\mathbf{A}$ are contained in r rows and s columns, where $r + s = \mu$, and that the order of the rows and columns is such that $\mathbf{A}$ contains, in the bottom left-hand corner, an $(m - r) \times (n - s)$ submatrix consisting entirely of 0s (see Fig. 27.4).

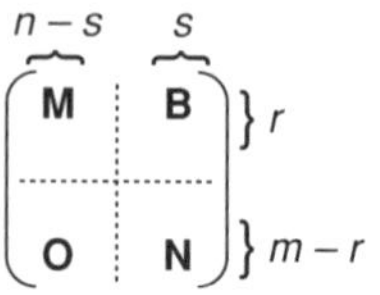

Fig. 27.4

If $i \leq r$, let S_i be the set of integers $j \leq n - s$ such that $a_{ij} = 1$. It is simple to check that the union of any k of the S_i contains at least k integers, and hence that the family $\mathcal{F} = (S_1, \ldots, S_r)$ has a transversal. It follows that the submatrix $\mathbf{M}$ of $\mathbf{A}$ contains a set of r 1s, no two of which lie in the same row or column. Similarly, the matrix $\mathbf{N}$ contains a set of s 1s with the same property. Hence $\mathbf{A}$ contains a set of $r + s$ 1s, no two of which lie in the same row or column. This shows that μ cannot exceed the term rank, as required. //

We have just proved the König–Egerváry theorem using Hall's theorem. It is even easier to prove Hall's theorem using the König–Egerváry theorem (see Exercise 27.5). It follows that these two theorems are, in some sense, equivalent. Later in this chapter we prove Menger's theorem and the max-flow min-cut theorem, each of which is also equivalent to Hall's theorem.

Common transversals

We conclude this section with a brief discussion of common transversals. If E is a non-empty finite set and $\mathcal{F} = (S_1, \ldots, S_m)$ and $\mathcal{G} = (T_1, \ldots, T_m)$ are two families of non-empty subsets of E, it is of interest to know when there exists a **common transversal** for $\mathcal{F}$ and $\mathcal{G}$ – that is, a set of m distinct elements of E that forms a transversal of both $\mathcal{F}$ and $\mathcal{G}$. In timetabling problems, for example, E may be the set of times at which lectures can be given, the sets S_i may be the times that m given professors are willing to lecture, and the sets T_i may be the times that m lecture rooms are available. Then the finding of a common transversal of $\mathcal{F}$ and $\mathcal{G}$ enables us to assign to each professor an available lecture room at a suitable time.

In Theorem 27.3 we can in fact give a necessary and sufficient condition for two families to have a common transversal. Note that it reduces to Hall's theorem if $T_j = E$ for $1 \leq j \leq m$.

THEOREM 27.3. *Let E be a non-empty finite set, and let $\mathcal{F} = (S_1, \ldots, S_m)$ and $\mathcal{G} = (T_1, \ldots, T_m)$ be two families of non-empty subsets of E. Then $\mathcal{F}$ and $\mathcal{G}$ have a common transversal if and only if, for all subsets A and B of $\{1, 2, \ldots, m\}$,*

$$P = \left| \left(\bigcup_{i \in A} S_i \right) \cap \left(\bigcup_{j \in B} T_j \right) \right| \geq |A| + |B| - m.$$

Sketch of proof. Consider the family $X = \{X_i\}$ of subsets of $E \cup \{1, \ldots, m\}$, assuming that E and $\{1, \ldots, m\}$ are disjoint, where the indexing set is also $E \cup \{1, \ldots, m\}$ and where $X_i = S_i$ if $i \in \{1, \ldots, m\}$ and $X_i = \{i\} \cup \{j : j \in T_j\}$ if $i \in E$. It is not difficult to verify that $\mathcal{F}$ and $\mathcal{G}$ have a common transversal if and only if X has a transversal. The result follows on applying Hall's theorem to the family X. //

It is not known when there exists a common transversal for three families of subsets of a set, and the problem of finding such conditions seems to be difficult. Many attempts to solve this problem use matroid theory. In fact, as we see in the next chapter, several results in transversal theory, such as those of Exercise 26.10 and Theorem 27.3, become much simpler when looked at from this viewpoint. Further results in transversal theory may also be found in Bryant and Perfect [25].

Exercises 27

27.1^s Give an example of a 5×8 latin rectangle and a 6×6 latin square.

27.2^s Find two ways of completing the following latin rectangle to a 5×5 latin square:
$$\begin{pmatrix} 1 & 2 & 3 & 4 & 5 \\ 5 & 3 & 1 & 2 & 4 \end{pmatrix}.$$

27.3   (i)   Use the result of Exercise 25.7 to prove that, if $m < n$, then an $m \times n$ latin rectangle can be extended to an $(m + 1) \times n$ latin rectangle in at least $(n - m)!$ ways.
(ii)   Deduce that the number of $n \times n$ latin squares is at least $n!(n-1)! \ldots 1!$.

27.4^s Verify the König–Egerváry theorem for the following matrices:
$$\begin{pmatrix} 0 & 0 & 1 & 0 & 1 \\ 1 & 0 & 1 & 1 & 1 \\ 0 & 1 & 1 & 0 & 0 \\ 0 & 0 & 0 & 0 & 1 \end{pmatrix} \quad \text{and} \quad \begin{pmatrix} 0 & 1 & 1 & 0 & 1 \\ 1 & 0 & 1 & 0 & 0 \\ 0 & 1 & 1 & 1 & 1 \\ 1 & 1 & 0 & 0 & 1 \end{pmatrix}.$$

27.5*   By regarding a $(0, 1)$-matrix as the incidence matrix of a family of subsets, show how the König–Egerváry theorem can be used to prove Hall's theorem.

27.6^s Let $E = \{a, b, c, d, e\}, \mathcal{F} = (\{a, c, e\}, \{a, b\}, \{c, d\})$ and $\mathcal{G} = (\{d\}, \{a, e\}, \{a, b, d\})$.
(i)   Find a common transversal of $\mathcal{F}$ and $\mathcal{G}$.
(ii)   Verify the condition of Theorem 27.3.

27.7   Repeat Exercise 27.6 for the families $\mathcal{F} = (\{a, b, d\}, \{c, e\}, \{a, e\})$ and $\mathcal{G} = (\{c, d\}, \{b\}, \{b, c, e\})$.

27.8*   Let G be a finite group and H be a subgroup of G. Use Theorem 27.3 to show that, if
$$G = x_1H \cup x_2H \cup \cdots \cup x_mH = Hy_1 \cup Hy_2 \cup \cdots \cup Hy_m$$
are left and right coset decompositions of G with respect to H, then there exist elements $z_1, \ldots, z_m$ in G such that
$$G = z_1H \cup z_2H \cup \cdots \cup z_mH = Hz_1 \cup Hz_2 \cup \cdots \cup Hz_m.$$

28 Menger's theorem

We now discuss a theorem that is closely related to Hall's theorem and has far-reaching practical applications. This theorem, due to K. Menger, concerns the number of paths connecting two given vertices v and w in a graph G. We may ask for the maximum number of paths from v to w, no two of which have an *edge* in common – such paths are called **edge-disjoint paths**. Alternatively, we may ask for the maximum number of paths from v to w, no two of which have a *vertex* in common, except, of course, v and w – these are called **vertex-disjoint paths**. For example, in the graph of Fig. 28.1, there are four edge-disjoint paths and two vertex-disjoint ones.

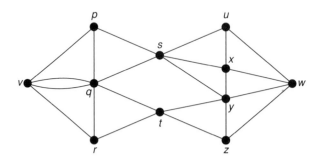

Fig. 28.1

In order to investigate these problems, we need some further definitions. We shall assume that G is a connected graph and that v and w are distinct vertices of G. A **vw-disconnecting set** of G is a set E of edges of G such that each path from v to w includes an edge of E; note that a vw-disconnecting set is a disconnecting set of G. Similarly, a **vw-separating set** of G is a set S of vertices, other than v or w, such that each path from v to w passes through a vertex of S. In Fig. 28.1, the sets $E_1 = \{ps, qs, ty, tz\}$ and $E_2 = \{uw, xw, yw, zw\}$ are vw-disconnecting sets, and $V_1 = \{s, t\}$ and $V_2 = \{p, q, y, z\}$ are vw-separating sets.

In order to count the edge-disjoint paths from v to w, we note first that, if E is a vw-disconnecting set with k edges, then the number of edge-disjoint paths cannot exceed k, since otherwise some edge in E would be included in more than one path. If E is a vw-disconnecting set of minimum possible size, then the number of edge-disjoint paths is actually equal to k, and there is exactly one edge of E in each such path. This result is known as the edge form of **Menger's theorem**; it was first proved in this form by Ford and Fulkerson in 1955.

> THEOREM 28.1. *The maximum number of edge-disjoint paths connecting two distinct vertices v and w of a connected graph is equal to the minimum number of edges in a vw-disconnecting set.*

Remark. Our proof is non-constructive, in that it does not provide us with a systematic way of obtaining k edge-disjoint paths, or even of finding the value of k. An algorithm that can be used to solve these problems is given in the next section.

Proof. As we have just pointed out, the maximum number of edge-disjoint paths connecting v and w cannot exceed the minimum number of edges in a vw-disconnecting set. We use induction on the number of edges of the graph G to prove that these numbers are equal. Suppose that the number of edges of G is m, and that the theorem is true for all graphs with fewer than m edges. There are two cases to consider.

(i) Suppose first that there exists a vw-disconnecting set E of minimum size k, such that not all of its edges are incident to v, and not all are incident to w. For example, in Fig. 28.1, the above set E_1 is such a vw-disconnecting set. The removal from G of the edges in E leaves two disjoint subgraphs V and W containing v and w, respectively. We now define two new graphs G_1 and G_2 as follows: G_1 is obtained from G by contracting every edge of V (that is, by shrinking V down to v), and G_2 is obtained by similarly contracting every edge of W; the graphs G_1 and G_2 obtained from Fig. 28.1 are shown in Fig. 28.2, with dashed lines denoting the edges of E_1. Since G_1 and G_2 have fewer edges than G, and since E is a vw-disconnecting set of minimum size for both G_1 and G_2, the induction hypothesis gives us k edge-disjoint paths in G_1 from v to w, and similarly for G_2. The required k edge-disjoint paths in G are obtained by combining these paths in the obvious way.

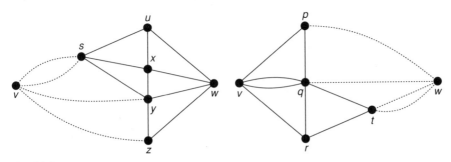

Fig. 28.2

(ii) Now suppose that each vw-disconnecting set of minimum size k consists only of edges that are all incident to v or all incident to w; for example, in Fig. 28.1, the set E_2 is such a vw-disconnecting set. We can assume without loss of generality that each edge of G is contained in a vw-disconnecting set of size k, since otherwise its removal would not affect the value of k and we could use the induction hypothesis to obtain k edge-disjoint paths. If P is any path from v to w, then P must consist of either one or two edges, and can thus contain at most one edge of any vw-disconnecting set of size k. By removing from G the edges of P, we obtain a graph with at least $k - 1$ edge-disjoint paths, by the induction hypothesis. These paths, together with P, give the required k paths in G. //

We turn now to the other problem mentioned at the beginning of the section – to find the number of vertex-disjoint paths from v to w. It was actually this problem that Menger himself solved, although his name is usually given to both of Theorems 28.1 and 28.2. The proof of Theorem 28.1 goes through with only minor changes, mainly involving the replacement of such terms as 'edge-disjoint' and 'incident' by 'vertex-disjoint' and 'adjacent'. We now state the vertex form of Menger's theorem – its proof is omitted.

THEOREM 28.2 (Menger, 1927). *The maximum number of vertex-disjoint paths connecting two distinct non-adjacent vertices v and w of a graph is equal to the minimum number of vertices in a vw-separating set.*

Using Theorems 28.1 and 28.2, we immediately deduce the following necessary and sufficient conditions for a graph to be *k*-connected and *k*-edge-connected:

COROLLARY 28.3. *A graph G is k-edge-connected if and only if any two distinct vertices of G are connected by at least k edge-disjoint paths.*

COROLLARY 28.4. *A graph G with at least k + 1 vertices is k-connected if and only if any two distinct vertices of G are connected by at least k vertex-disjoint paths.*

The above discussion can be modified to give the number of arc-disjoint paths from a vertex *v* to a vertex *w* in a digraph; in this case, we can take *v* to be a source and *w* to be a sink. The resulting theorem is similar to Theorem 28.1, and the proof goes through almost word for word. Note that, in a digraph, a *vw*-disconnecting set is a set *A* of arcs such that each path from *v* to *w* includes an arc in *A*.

THEOREM 28.5. *The maximum number of arc-disjoint paths from a vertex v to a vertex w in a digraph is equal to the minimum number of arcs in a vw-disconnecting set.*

For example, if the digraph is as shown in Fig. 28.3, then there are six arc-disjoint paths from *v* to *w*. A corresponding *vw*-disconnecting set consists of the arcs *vz, xz, yz* (twice) and *xw* (twice).

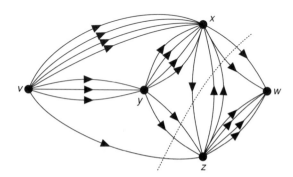

Fig. 28.3

These diagrams become very cumbersome as the number of arcs joining pairs of vertices increases. To overcome this, we draw just one arc and write next to it the number of arcs there should be (see Fig. 28.4). This seemingly innocent remark is fundamental to the study of network flows, which we discuss in the next section.

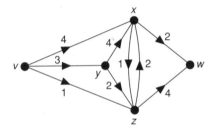

Fig. 28.4

We end this section by deducing Hall's theorem from Menger's theorem. We prove the version of Hall's theorem that appears in Corollary 25.2.

THEOREM 28.6. *Menger's theorem implies Hall's theorem.*

Proof. Let $G = G(V_1, V_2)$ be a bipartite graph. We must prove that, if $|A| \leq |\varphi(A)|$ for each subset A of V_1, then there is a complete matching from V_1 to V_2. To do this, we apply the vertex form of Menger's theorem (Theorem 28.2) to the graph obtained by adjoining to G a vertex v adjacent to every vertex in V_1 and a vertex w adjacent to every vertex in V_2 (see Fig. 28.5).

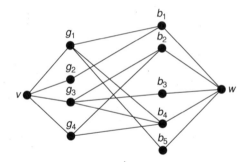

Fig. 28.5

Since a complete matching from V_1 to V_2 exists if and only if the number of vertex-disjoint paths from v to w is equal to the number of vertices in V_1 (= k, say), it is enough to show that every vw-separating set has at least k vertices. So, let S be a vw-separating set consisting of a subset A of V_1 and a subset B of V_2. Since $A \cup B$ is a vw-separating set, no edge can join a vertex of $V_1 - A$ to a vertex of $V_2 - B$, and hence $\varphi(V_1 - A) \subseteq B$. It follows that

$$|V_1 - A| \leq |\varphi(V_1 - A)| \leq |B|,$$

and so $|S| = |A| + |B| \geq |V_1| = k$, as required. //

Exercises 28

28.1[s] Verify Theorems 28.1 and 28.2 for each graph in Fig. 28.6.

28.2[s] Verify Theorems 28.1 and 28.2 for the Petersen graph in the two cases:
    (i)   when v and w are adjacent vertices;
    (ii) when v and w are not adjacent.

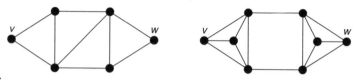

Fig. 28.6

28.3   Prove Theorem 28.2 in detail.

28.4^s   Verify Corollary 28.3 for each of the following graphs:
(i) W_5; (ii) $K_{3,4}$; (iii) Q_3.

28.5   Verify Corollary 28.4 for each of the following graphs:
(i) $K_{3,5}$; (ii) $K_{3,3,3}$; (iii) the graph of the octahedron.

28.6   Verify Theorem 28.5 for each digraph in Fig. 28.7.

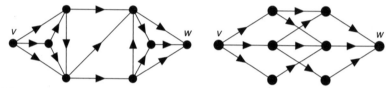

Fig. 28.7

29   Network flows

Our society today is largely governed by networks – transportation, communication, etc. – and the mathematical analysis of such networks has become of fundamental importance. In this section we indicate that network analysis is essentially the study of digraphs.

A computer manufacturer wishes to send several computers to a given market. There are various channels through which the boxes can be sent, as shown in Fig. 29.1, with v representing the manufacturer and w the market. The number next to each arc refers to the maximum load that can pass through the corresponding channel. The manufacturer wishes to find the maximum number of boxes that can be sent through the network without exceeding the permitted capacity of any channel.

Figure 29.1 can also describe other situations. For example, if each arc represents a one-way street and the number next to each arc is the maximum possible flow of traffic along that street, in vehicles per hour, then we may ask for the greatest possible number of vehicles that can travel from v to w in one hour. Alternatively, if the diagram

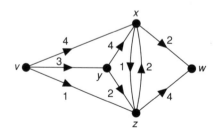

Fig. 29.1

depicts an electrical network, we may ask for the maximum current that can safely be passed through the network, given the currents at which individual wires burn out.

Using these examples as motivation, we define a **network** N to be a weighted digraph – that is, a digraph to each arc a of which is assigned a non-negative real number $\psi(a)$ called its **capacity**. The **out-degree** outdeg(x) of a vertex x is the sum of the capacities of the arcs of the form xz, and the **in-degree** indeg(x) is similarly defined. For example, in the network of Fig. 29.1, outdeg$(v) = 8$ and indeg$(x) = 10$. Note that the handshaking dilemma now takes the form: *the sum of the out-degrees of the vertices of a network is equal to the sum of the in-degrees.*

We assume that the digraph has exactly one source v and one sink w. The general case of several sources and sinks, corresponding to more than one manufacturer and market, is easily reduced to this special case (see Exercise 29.5).

A **flow** in a network is a function φ that assigns to each arc a a non-negative real number $\varphi(a)$, called the **flow in** a, in such a way that

(i) for each arc a, $\varphi(a) \leq \psi(a)$;
(ii) the out-degree and in-degree of each vertex, other than v or w, are equal.

Thus, the flow in any arc cannot exceed its capacity and the 'total flow' into each vertex, other than v or w, is equal to the 'total flow' out of it. Figure 29.2 gives a possible flow for the network of Fig. 29.1. Another flow is the **zero flow** in which the flow in every arc is 0; any other flow is a **non-zero flow**. An arc a for which $\varphi(a) = \psi(a)$ is called **saturated**. In Fig. 29.2, the arcs vz, xz, yz, xw and zw are saturated, and the remaining arcs are **unsaturated**.

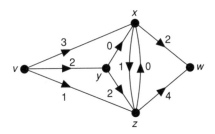

Fig. 29.2

It follows from the handshaking dilemma that the sum of the flows in the arcs out of v is equal to the sum of the flows in the arcs into w; this sum is called the **value of the flow**. Prompted by our earlier examples, we are mainly interested in flows whose value is as large as possible – the **maximum flows**. You can easily check that the flow of Fig. 29.2 is a maximum flow for the network of Fig. 29.1, and that its value is 6. Although a network can have several different maximum flows, their values must be equal.

The study of maximum flows in a network is closely tied up with the concept of a **cut**, which is a set A of arcs such that each path from v to w includes an arc in A. Thus, a cut in a network is a vw-disconnecting set in the corresponding digraph D. The **capacity of a cut** is the sum of the capacities of the arcs in the cut. We are concerned mainly with those cuts whose capacity is as small as possible, the so-called **minimum cuts**. In Fig. 29.3, a minimum cut consists of the arcs vz, xz, yz and xw, but not the arc zx; the capacity of this cut is $2 + 2 + 2 = 6$.

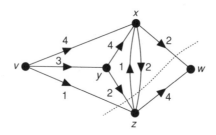

Fig. 29.3

Note that the value of any flow cannot exceed the capacity of any cut, and so the value of any *maximum* flow cannot exceed the capacity of any *minimum* cut. It turns out that these last two numbers are always equal. This famous result, known as the **max-flow min-cut theorem**, was first proved by Ford and Fulkerson in 1955. We present two proofs. The first shows that the max-flow min-cut theorem is equivalent to Menger's theorem; the second is a direct proof.

> **THEOREM 29.1 (Max-flow min-cut theorem).** *In any network, the value of any maximum flow is equal to the capacity of any minimum cut.*

Remark. When applying this theorem, it is often simplest to find a flow and a cut such that the value of the flow equals the capacity of the cut. It follows from the theorem that the flow must be a maximum flow and that the cut must be a minimum cut. If all the capacities are integers, then the value of a maximum flow is also an integer; this turns out to be useful in certain applications of network flows.

First proof. Suppose first that the capacity of each arc is an integer. Then the network can be regarded as a digraph D whose capacities represent the number of arcs connecting the various vertices (as in Figs. 28.3 and 28.4). The value of a maximum flow is the total number of arc-disjoint paths from v to w in D, and the capacity of a minimum cut is the minimum number of arcs in a vw-disconnecting set of D. The result now follows from Theorem 28.5.

The extension of this result to networks in which the capacities are rational numbers is effected by multiplying these capacities by a suitable integer d to make them integers. We then have the case described above, and the result follows on dividing by d.

Finally, if some capacities are irrational, then we approximate them as closely as we please by rationals and use the above result. By choosing these rationals carefully, we can ensure that the value of any maximum flow and the capacity of any minimum cut are altered by an amount that is as small as we wish. Note that, in practical examples, irrational capacities rarely occur, since the capacities are usually given in decimal form. //

Second proof. Since the value of any maximum flow cannot exceed the capacity of any minimum cut, it is sufficient to prove the existence of a cut whose capacity is equal to the value of a given maximum flow.

Let φ be a maximum flow. We define two sets V and W of vertices of the network as follows. If G is the underlying graph of the network, then a vertex z is contained in V if and only if there exists in G a path $v = v_0 \rightarrow v_1 \rightarrow v_2 \rightarrow \cdots \rightarrow v_{m-1} \rightarrow v_m = z$, such that each edge v_iv_{i+1} corresponds either to an unsaturated arc v_iv_{i+1}, or to an arc $v_{i+1}v_i$ that carries a non-zero flow. The set W consists of all those vertices that do not lie in V. For example, in Fig. 29.2, the set V consists of the vertices v, x and y, and the set W consists of the vertices z and w.

Clearly, v is contained in V. We now show that W contains the vertex w. If this is not so, then w lies in V, and hence there exists in G a path $v \rightarrow v_1 \rightarrow v_2 \rightarrow \cdots \rightarrow v_{m-1} \rightarrow w$ of the above type. We now choose a positive number ε that does not exceed the amount needed to saturate any unsaturated arc v_iv_{i+1}, and does not exceed the flow in any arc $v_{i+1}v_i$ that carries a non-zero flow. It is now easy to see that, if we increase by ε the flow in all arcs of the first type and decrease by ε the flow in all arcs of the second type, then we increase the value of the flow by ε, contradicting our assumption that φ is a maximum flow. It follows that w lies in W.

To complete the argument, we let E be the set of all arcs of the form xz, where x is in V and z is in W. Clearly E is a cut. Moreover, each arc xz of E is saturated and each arc zx carries a zero flow, since otherwise z would also be an element of V. It follows that the capacity of E must equal the value of φ, and that E is the required minimum cut. //

The max-flow min-cut theorem provides a useful check on the maximality or otherwise of a given flow, as long as the network is fairly simple. In practice, the networks we have to deal with are large and complicated, and it is usually difficult to find a maximum flow by inspection. Most methods for finding a maximum flow involve determining **flow-augmenting paths** from v to w. These are paths consisting entirely of unsaturated arcs xz and arcs zx that carry a non-zero flow. For example, consider the network of Fig. 29.4.

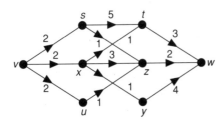

Fig. 29.4

Starting with the zero flow, we can construct the flow-augmenting paths $v \rightarrow s \rightarrow t \rightarrow w$ along which the value of the flow can be increased by 2, $v \rightarrow x \rightarrow z \rightarrow w$ along which the value of the flow can be increased by 2, and $v \rightarrow u \rightarrow z \rightarrow x \rightarrow y \rightarrow w$ along which the value of the flow can be increased by 1. The resulting flow of value 5 is as follows.

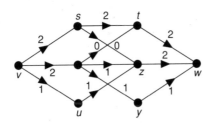

Fig. 29.5

Since the network has a cut of capacity 5, the above flow is a maximum flow and the cut is a minimum cut.

In this section we have been able only to scratch the surface of this very diverse and important subject. If you wish to pursue these topics, see Lawler [17].

Exercises 29

29.1[s]   Consider the network of Fig. 29.6.
   (i)   List all the cuts in this network, and find a minimum cut.
   (ii)   Find a maximum flow, and verify the max-flow min-cut theorem.

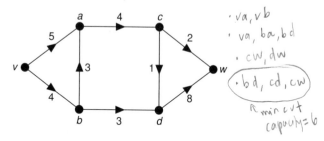

· va, vb

· va, ba, bd

· cw, dw

·bd, cd, cw

↞ min cut

capouM=6

Fig. 29.6

29.2   Repeat Exercise 29.1 for the network of Fig. 29.7.

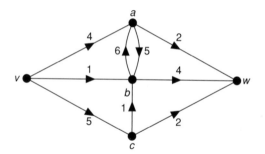

Fig. 29.7

29.3[s]   Verify the max-flow min-cut theorem for the network of Fig. 22.8.

29.4   Find a flow with value 20 in the network of Fig. 29.8. Is it a maximum flow?

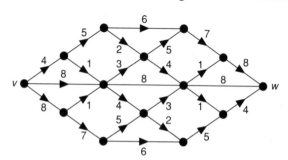

Fig. 29.8

29.5[s]   (i)   Show how the analysis of the flows in a network with several sources and sinks can be reduced to the standard case by the addition of a new 'source vertex' and 'sink vertex'.
   (ii)   Illustrate your answer to part (i) with reference to the network in Fig. 29.9.

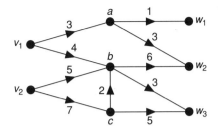

Fig. 29.9

29.6 (i) How would you reduce to the standard case a network problem in which some arcs are replaced by edges with a flow in either direction, and some vertices are assigned 'capacities' indicating the maximum flow permitted through those vertices?

(ii) Illustrate your answers to part (i) with reference to the network in Fig. 29.10.

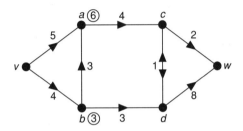

Fig. 29.10

29.7* Show how the max-flow min-cut theorem can be used to prove
(i) Hall's theorem;
(ii) Theorem 27.3 on common transversals.

Chapter 9

Matroids

The first of earthly blessings, independence.
Edward Gibbon

In this chapter we investigate an unexpected similarity between certain results in graph theory and their analogues in transversal theory, as illustrated by Exercises 9.11 and 26.8, or Exercises 9.12 and 26.9. To do this, we introduce the idea of a matroid, first studied in 1935 in a pioneering paper by Hassler Whitney. As we shall see, a matroid is a set with an 'independence structure' defined on it, where the notion of independence generalizes that of independence in graphs, as defined in Exercise 5.13, and of linear independence in vector spaces. The link with transversal theory is provided by Exercise 26.8.

In Section 32 we define duality in matroids in such a way as to explain the similarity between the properties of cycles and cutsets in a graph. The unintuitive definition of an abstract dual of a graph in Section 15 then arises as a natural consequence of matroid duality. In the final section, we show how matroids can be used to give 'easy' proofs of results in transversal theory, and we conclude with matroid proofs of two deep results in graph theory. Throughout this chapter we state results without proof, where convenient. The omitted proofs may be found in Oxley [34] or Welsh [37].

30 Introduction to matroids

In Section 9 we defined a spanning tree in a connected graph G to be a connected subgraph of G that contains no cycles and includes every vertex of G. Note that a spanning tree cannot contain another spanning tree as a proper subgraph. We also saw in Exercise 9.11 that, if B_1 and B_2 are spanning trees of G and e is an edge of B_1, then there is an edge f in B_2 such that $(B_1 - \{e\}) \cup \{f\}$ (the graph obtained from B_1 on replacing e by f) is also a spanning tree of G.

Analogous results hold in the theory of vector spaces and in transversal theory. If V is a vector space and if B_1 and B_2 are bases of V and e is an element of B_1, then we can find an element f of B_2 such that $(B_1 - \{e\}) \cup \{f\}$ is also a basis of V. The corresponding result in transversal theory appears in Exercise 26.8.

Using these examples as motivation, we now give our first definition of a matroid.

A **matroid** M consists of a non-empty finite set E and a non-empty collection $\mathcal{B}$ of subsets of E, called **bases**, satisfying the following properties:

$\mathcal{B}$(i)   no base properly contains another base;

$\mathcal{B}$(ii)   if B_1 and B_2 are bases and if e is any element of B_1, then there is an element f of B_2 such that $(B_1 - \{e\}) \cup \{f\}$ is also a base.

By repeatedly using property $\mathcal{B}$(ii), we can easily show that any two bases of a matroid M have the same number of elements (see Exercise 30.5). This number is called the **rank** of M.

As we indicated above, a matroid can be associated with any graph G by letting E be the set of edges of G and taking as bases the edges of the spanning forests of G. For reasons that will appear later, this matroid is called the **cycle matroid** of G and is denoted by $M(G)$. Similarly, if E is a finite set of vectors in a vector space V, then we can define a matroid on E by taking as bases all linearly independent subsets of E that span the same subspace as E. A matroid obtained in this way is called a **vector matroid**. We consider such matroids later.

A subset of E is **independent** if it is contained in some base of the matroid M. For a vector matroid, a subset of E is independent whenever its elements are linearly independent as vectors in the vector space. For the cycle matroid $M(G)$ of a graph G, the independent sets are those sets of edges of G that contain no cycle – that is, the edge sets of the forests contained in G.

Since the bases of M are the maximal independent sets (that is, those independent sets contained in no larger independent set), a matroid is uniquely defined by specifying its independent sets. It is natural to ask whether there is a simple definition of a matroid in terms of its independent sets. One such definition is as follows. A proof of the equivalence of this definition and the one above is given in Welsh [37].

A **matroid** M consists of a non-empty finite set E and a non-empty collection I of subsets of E (called **independent sets**) satisfying the following properties:

I(i)   any subset of an independent set is independent;

I(ii)   if I and J are independent sets with $|J| > |I|$, then there is an element e, contained in J but not in I, such that $I \cup \{e\}$ is independent.

With this definition, a **base** is defined to be a maximal independent set. Property I(ii) can then be used repeatedly to show that any independent set can be extended to a base.

If $M = (E, I)$ is a matroid defined in terms of its independent sets, then a subset of E is **dependent** if it is not independent, and a minimal dependent set is called a **cycle**. If $M(G)$ is the cycle matroid of a graph G, then the cycles of $M(G)$ are precisely the cycles of G. Since a subset of E is independent if and only if it contains no cycles, a matroid can be defined in terms of its cycles. One such definition, generalizing to matroids the result of Exercise 5.11, is given in Exercise 30.7.

Before proceeding to some examples of matroids, we give one further definition of a matroid. This definition, in terms of a rank function r, is essentially the one given by Whitney.

If $M = (E, I)$ is a matroid defined in terms of its independent sets, and if A is a subset of E, then the **rank** of A, denoted by $r(A)$, is the size of the largest independent set contained in A. Note that the previously defined rank of M is equal to $r(E)$.

Since a subset A of E is independent if and only if $r(A) = |A|$, we can define a matroid in terms of its rank function.

THEOREM 30.1. *A matroid consists of a non-empty finite set E and an integer-valued function r defined on the set of subsets of E, satisfying:*

   r(i) $0 \le r(A) \le |A|$, for each subset A of E;

   r(ii) if $A \subseteq B \subseteq E$, then $r(A) \le r(B)$;

   r(iii) for any $A, B \subseteq E, r(A \cup B) + r(A \cap B) \le r(A) + r(B)$.

Remark. This theorem extends to matroids the results of Exercises 9.12 and 26.9.

Proof. We assume first that M is a matroid defined in terms of its independent sets. We wish to prove properties r(i)–r(iii). Clearly, properties r(i) and r(ii) are trivial. To prove r(iii), we let X be a base (a maximal independent subset) of $A \cap B$. Since X is an independent subset of A, X can be extended to a base Y of A, and then (in a similar way) to a base Z of $A \cup B$. Since $X \cup (Z - Y)$ is clearly an independent subset of B, it follows that

$$\begin{aligned} r(B) \; &\ge r(X \cup (Z - Y)) \\ &= |X| + |Z| - |Y| \\ &= r(A \cap B) + r(A \cup B) - r(A), \end{aligned}$$

as required.

Conversely, let $M = (E, r)$ be a matroid defined in terms of a rank function r, and define a subset A of E to be independent if and only if $r(A) = |A|$. It is then simple to prove property I(i). To prove I(ii), let I and J be independent sets with $|J| > |I|$, and suppose that $r(I \cup \{e\}) = k$ for each element e that lies in J but not in I. If e and f are two such elements, then

$$r(I \cup \{e\} \cup \{f\}) \le r(I \cup \{e\}) + r(I \cup \{f\}) - r(I) = k.$$

It follows that $r(I \cup \{e\} \cup \{f\}) = k$. We now continue this procedure, adding one new element of J at a time. Since at each stage the rank k, we conclude that $r(I \cup J) = k$, and hence (by r(ii)) that $r(J) \le k$, which is a contradiction. It follows that there exists an element f that lies in J but not in I, such that $r(I \cup \{f\}) = k + 1$. //

We conclude this section with two further definitions. A **loop** of a matroid M is an element e of E satisfying $r(\{e\}) = 0$, and a pair of **parallel elements** of M is a pair $\{e, f\}$ of elements of E that satisfy $r(\{e, f\}) = 1$ and are not loops. Note that, if M is the cycle matroid of a graph G, then the loops and parallel elements of M correspond to loops and multiple edges of G.

Exercises 30

30.1^s   Let $E = \{a, b, c, d, e\}$. Find matroids on E for which
   (i) E is the only base;
   (ii) the empty set $\varnothing$ is the only base;
   (iii) the bases are those subsets of E containing exactly three elements.
   For each matroid, write down the independent sets, the cycles (if there are any) and the rank function.
   (This question is answered in the next section.)

30.2^s   Let G_1 and G_2 be the graphs shown in Fig. 30.1. Write down the bases, cycles and independent sets of the cycle matroids $M(G_1)$ and $M(G_2)$.

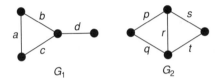

Fig. 30.1

30.3 Let M be the matroid on the set $E = \{a, b, c, d\}$ whose bases are $\{a, b\}$, $\{a, c\}$, $\{a, d\}$, $\{b, c\}$, $\{b, d\}$ and $\{c, d\}$. Write down the cycles of M, and deduce that there is no graph G with M as its cycle matroid.

30.4ˢ Let $E = \{1, 2, 3, 4, 5, 6\}$ and $\mathcal{F} = (S_1, S_2, S_3, S_4, S_5)$, where

$$S_1 = S_2 = \{1, 2\}, S_3 = S_4 = \{2, 3\}, S_5 = \{1, 4, 5, 6\}.$$

   (i)   Write down the partial transversals of $\mathcal{F}$ and check that they form the independent sets of a matroid on E.

   (ii)   Write down the bases and cycles of this matroid.

30.5 Use properties $\mathcal{B}(i)$ and $\mathcal{B}(ii)$ to prove that

   (i)   any two bases of a matroid on a set E have the same number of elements;

   (ii)   if $A \subseteq E$, then any two maximal independent subsets of A have the same number of elements.

30.6 Show how the definition of a fundamental system of cycles in a graph can be extended to matroids.

30.7* Show that a matroid M can be defined as follows:

a **matroid** consists of a non-empty finite set E, and a collection C of non-empty subsets of E (called **cycles**) satisfying the following properties:

$C(i)$ no cycle properly contains another cycle;

$C(ii)$ if C_1 and C_2 are two distinct cycles each containing an element e, then there exists a cycle in $C_1 \cup C_2$ that does not contain e.

30.8ˢ (i)   Use the result of Exercise 5.11 to show that the cutsets of a graph satisfy conditions $C(i)$ and $C(ii)$ of Exercise 30.7.

   (ii)   Write down the bases of the corresponding matroids for the graphs of Fig. 30.1.

30.9* State and prove a matroid analogue of the greedy algorithm (Theorem 11.1).

31 *Examples of matroids*

In this section we examine several important types of matroid.

Trivial matroids

Given any non-empty finite set E, we can define on it a matroid whose only independent set is the empty set $\varnothing$. This matroid is the **trivial matroid** on E, and has rank 0.

Discrete matroids

At the other extreme is the **discrete matroid** on E, in which every subset of E is independent. Note that the discrete matroid on E has only one base, E itself, and that the rank of any subset A is the number of elements in A.

Uniform matroids

The previous examples are special cases of the **k-uniform matroid** on E, whose bases are those subsets of E with exactly k elements; the trivial matroid on E is 0-uniform and the discrete matroid is $|E|$-uniform. Note that the independent sets are those subsets of E with not more than k elements, and that the rank of any subset A is either $|A|$ or k, whichever is smaller.

Before developing the examples described in the previous section, we define two matroids M_1 and M_2 to be **isomorphic** if there is a one–one correspondence between their underlying sets E_1 and E_2 that preserves independence. Thus, a set of elements of E_1 is independent in M_1 if and only if the corresponding set of elements of E_2 is independent in M_2. For example, the cycle matroids of the three graphs in Fig. 31.1 are all isomorphic. Note that, although matroid isomorphism preserves cycles, cutsets and the number of edges in a graph, it does not necessarily preserve connectedness, the number of vertices, or their degrees.

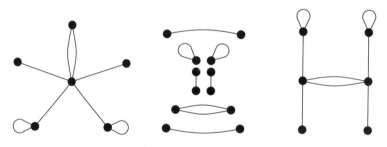

Fig. 31.1

Using this definition of isomorphism, we can define graphic, representable and transversal matroids.

Graphic matroids

As we saw in the previous section, we can define a matroid $M(G)$ on the set of edges of a graph G by taking the cycles of G as the cycles of the matroid. $M(G)$ is the **cycle matroid** of G and its rank function is the cutset rank ξ (see Exercise 9.12). It is natural to ask whether a given matroid M is the cycle matroid of some graph; in other words, does there exists a graph G such that M is isomorphic to $M(G)$? Such matroids are called **graphic matroids**, and a characterization of them is given in the next section. For example, the matroid M on the set $\{1, 2, 3\}$ whose bases are $\{1, 2\}$, and $\{1, 3\}$ is a graphic matroid isomorphic to the cycle matroid of the graph in Fig. 31.2. A simple example of a non-graphic matroid is the 2-uniform matroid on a set of four elements (see Exercise 30.3).

Fig. 31.2

Cographic matroids

Given a graph G, the cycle matroid $M(G)$ is not the only matroid that can be defined on the set of edges of G. Because of the similarity between the properties of cycles and of cutsets in a graph, we can construct a matroid by taking the *cutsets* of G as cycles of the matroid. We saw in Exercise 30.8 that this construction defines a matroid, and we call it the **cutset matroid** of G, denoted by $M^*(G)$. Note that a set of edges of G is independent if and only if it contains no cutset of G. We call a matroid M **cographic** if there exists a graph G such that M is isomorphic to $M^*(G)$. The reason for the name 'cographic' is given in the next section.

Planar matroids

A matroid that is both graphic and cographic is a **planar matroid**. In the next section, we indicate the connection between planar matroids and planar graphs.

Representable matroids

Since the definition of a matroid is partly motivated by linear independence in vector spaces, it is of interest to investigate those matroids that arise as vector matroids associated with some set of vectors in a vector space. Given a matroid M on a set E, we say that M is **representable over a field F** if there exist a vector space V over F and a map φ from E to V, such that a subset A of E is independent in M if and only if φ is one–one on A and $\varphi(A)$ is linearly independent in V. This amounts to saying that, if we ignore loops and parallel elements, then M is isomorphic to a vector matroid defined in some vector space over F. For convenience, we say that M is a **representable matroid** if there exists some field F such that M is representable over F.

It turns out that some matroids are representable over every field (the **regular matroids**), some are representable over no field, and some are representable only over a restricted class of fields. Of particular importance are the **binary matroids**, representable over the field of integers modulo 2. For example, if G is any graph, then its cycle matroid $M(G)$ is a binary matroid. To see this, we associate with each edge of G the corresponding row of the incidence matrix of G, regarded as a vector with components 0 or 1. Note that, if a set of edges of G forms a cycle, then the sum (modulo 2) of the corresponding vectors is 0.

A binary matroid that is neither graphic nor cographic is the Fano matroid, described at the end of this section.

Transversal matroids

Our next example provides the link between matroids and transversal theory. Recall from Exercises 26.8, 26.9 and 30.4 that, if E is a non-empty finite set and if $\mathcal{F} = (S_1, \ldots, S_m)$ is a family of non-empty subsets of E, then the partial transversals of $\mathcal{F}$ can be taken as the independent sets of a matroid on E, denoted by $M(\mathcal{F})$ or $M(S_1, \ldots, S_m)$. Any matroid obtained in this way (for suitable choices of E and $\mathcal{F}$) is a **transversal matroid**. For example, the above graphic matroid M is a transversal matroid on the set $\{1, 2, 3\}$, since its independent sets are the partial transversals of the family $\mathcal{F} = (S_1, S_2)$, where $S_1 = \{1\}$ and $S_2 = \{2, 3\}$. Note that the rank of a subset A of E is the size of the largest partial transversal contained in A. An example of a non-transversal matroid is given in Exercise 31.5.

It can be proved that every transversal matroid is representable over some field, but is binary if and only if it is graphic. Further results on transversal matroids are discussed in Section 33.

Restrictions and contractions

In graph theory we often investigate the properties of a graph by looking at its sub-graphs, or by considering the graph obtained by contracting some of its edges. We now define the corresponding notions for matroids.

If M is a matroid defined on a set E, and if A is a subset of E, then the **restriction** of M to A, denoted by $M \times A$, is the matroid whose cycles are precisely those cycles of M that are contained in A. Similarly, the **contraction** of M to A, denoted by $M \cdot A$, is the matroid whose cycles are the minimal members of the collection $\{C_i \cap A\}$, where C_i is a cycle of M. (A simpler definition is given in Exercise 32.7.) You can verify that these are indeed matroids, and that they correspond to the deletion and contraction of edges in a graph. A matroid obtained from M by restrictions and/or contractions is called a **minor** of M.

Bipartite and Eulerian matroids

We next define bipartite and Eulerian matroids. Since the usual definitions of bipartite and Eulerian graphs, given in Sections 3 and 6, are unsuitable for matroid generalization, we must use alternative characterizations of these graphs. In the case of bipartite graphs, we use Exercise 5.3, and define a **bipartite matroid** to be a matroid in which each cycle has an even number of elements. For Eulerian graphs we use Corollary 6.3 and define a matroid on a set E to be an **Eulerian matroid** if E can be written as a union of disjoint cycles. In the next section we see that Eulerian matroids and bipartite matroids are dual concepts, in a sense to be made precise, as we might expect from Exercise 15.9.

The Fano matroid

The **Fano matroid** F is the matroid defined on the set $E = \{1, 2, 3, 4, 5, 6, 7\}$, whose bases are all those subsets of E with three elements, except $\{1, 2, 4\}$, $\{2, 3, 5\}$, $\{3, 4,$

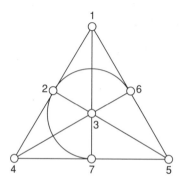

Fig. 31.3

6}, {4, 5, 7}, {5, 6, 1}, {6, 7, 2} and {7, 1, 3}. This matroid can be represented geometrically by Fig. 31.3; the bases are precisely those sets of three elements that do not
lie on a line. It can be shown that F is binary and Eulerian, but is not graphic,
cographic, transversal or regular.

Exercises 31

31.1[s]   Let $E = \{a, b\}$. Show that there are (up to isomorphism) exactly four matroids on E,
and list their bases, independent sets and cycles.

31.2   Let $E = \{a, b, c\}$. Show that there are (up to isomorphism) exactly eight matroids on E,
and list their bases, independent sets and cycles.

31.3*   Let E be a set of n elements. Show that, up to isomorphism,
   (i)   the number of matroids on E is at most 2^{2^n};
   (ii)   the number of transversal matroids on E is at most 2^{n^2}.

31.4[s]   Let G_1 and G_2 be the graphs of Fig. 30.1.
   (i)   Are $M(G_1)$ and $M(G_2)$ transversal matroids?
   (ii)   Are $M^*(G_1)$ and $M^*(G_2)$ transversal matroids?

31.5   Show that $M(K_4)$ is not a transversal matroid.

31.6[s]   Show that every uniform matroid is a transversal matroid.

31.7   Show that the graphic matroids $M(K_5)$ and $M(K_{3,3})$ are not cographic.

31.8[s]   Describe the cycles of the Fano matroid.

31.9   Let M be a matroid on a set E, and let $A \subseteq B \subseteq E$. Prove that
   (i) $(M \times B) \times A = M \times A$;
   (ii) $(M . B) . A = M . A$.

***31.10**   Prove that, if M satisfies any of the following properties, then so does any minor of M:
   (i) graphic; (ii) cographic; (iii) binary; (iv) regular.

32   Matroids and graphs

We now study duality in matroids. Our aim is to show how some of our earlier results
seem more natural when looked at in this light. We shall see, for example, that the
rather artificial definition of an abstract dual of a planar graph in Section 15 arises as a
direct consequence of the corresponding definition of a matroid dual. Indeed, not only
do concepts in matroid theory generalize their counterparts in graph theory, but they
sometimes simplify them as well.

   Recall that we can form a matroid $M^*(G)$ on the set of edges of a graph G by taking as cycles of $M^*(G)$ the cutsets of G. In view of Theorem 15.3, it is sensible to
define a matroid dual so as to make this matroid the dual of the cycle matroid $M(G)$.

   This is achieved as follows: if M is a matroid on a set E, defined in terms of its rank
function, we define the **dual matroid** M^* of M to be the matroid on E whose rank
function r^* is given by the expression

$$r^*(A) = |A| + r(E - A) - r(E), \text{ for } A \subseteq E.$$

We first verify that r^* is the rank function of a matroid on E.

THEOREM 32.1. $M^* = (E, r^*)$ *is a matroid on E.*

Proof. We verify the properties r(i) and r(iii) of Section 30, for the function r^*; the proof of r(ii) is equally straightforward (see Exercise 32.3).

To prove r(i), note that $r(E - A) \leq r(E)$, and hence that $r^*(A) \leq |A|$. Also, by r(iii) applied to the function r, we have

$$r(E) + r(\emptyset) \leq r(A) + r(E - A),$$

and hence

$$r(E) - r(E - A) \leq r(A) \leq |A|.$$

It follows immediately that $r^*(A) \geq 0$.

To prove r(iii), we have, for any $A, B \subseteq E$,

$$\begin{aligned}
r^*(A \cup B) + r^*(A \cap B) &= |A \cup B| + |A \cap B| + r(E - (A \cup B)) \\
&\quad + r(E - (A \cap B)) - 2r(E) \\
&= |A| + |B| + r((E - A) \cap (E - B)) \\
&\quad + r((E - A) \cup (E - B)) - 2r(E) \\
&\leq |A| + |B| + r(E - A) + r(E - B) - 2r(E) \\
&\qquad\qquad\qquad \text{(by } r\text{(iii), applied to } r\text{)} \\
&= r^*(A) + r^*(B),
\end{aligned}$$

as required. //

Although the above definition may seem contrived, it turns out that the bases of M^* can be described very simply in terms of those of M, as we now show.

THEOREM 32.2. *The bases of M^* are precisely the complements of the bases of M.*

Remark. This result is often used to define M^*.

Proof. We show that, if B^* is a base of M^*, then $E - B^*$ is a base of M; the converse result is obtained by reversing the argument. Since B^* is independent in M^*, $|B^*| = r^*(B^*)$, and hence $r(E - B^*) = r(E)$. It remains only to prove that $E - B^*$ is independent in M. But this follows immediately from the fact that $r^*(B^*) = r^*(E)$, on using the above expression for r^*. //

It follows from the above definition that, unlike the duality of planar graphs, every matroid has a dual and this dual is unique. It also follows immediately from Theorem 32.2 that the double-dual M^{**} is equal to M. As we shall see, this trivial result is the natural generalization to matroids of the non-trivial Theorems 15.2 and 15.5.

We now show that the cutset matroid $M^*(G)$ of a graph G is the dual of the cycle matroid $M(G)$.

THEOREM 32.3. *If G is a graph, then* $M^*(G) = (M(G))^*$.

Proof. Since the cycles of $M^*(G)$ are the cutsets of G, we must check that C^* is a cycle of $(M(G))^*$ if and only if C^* is a cutset of G.

Suppose first that C^* is a cutset of G. If C^* is independent in $(M(G))^*$, then C^* can be extended to a base B^* of $(M(G))^*$, and so $C^* \cap (E - B^*)$ is empty. This contradicts the result of Theorem 9.3(i), since $E - B^*$ is a spanning forest of G. Thus, C^* is a dependent set in $(M(G))^*$, and therefore contains a cycle of $(M(G))^*$.

If, on the other hand, D^* is a cycle of $(M(G))^*$, then D^* is not contained in any base of $(M(G))^*$. It follows that D^* intersects every base of $M(G)$ – that is, every spanning forest of G. Thus, by the result of Exercise 9.10(i), D^* contains a cutset. The result follows. //

Before proceeding further, we introduce some more terminology. We say that elements of a matroid M form a **cocycle** of M if they form a cycle of M^*. Note that, in view of Theorem 32.3, the cocycles of the cycle matroid of a graph G are precisely the cutsets of G. We similarly define a **cobase** of M to be a base of M^*, with corresponding definitions for **corank, co-independent set,** etc. We also say that a matroid M **is cographic** if and only if its dual M^* is graphic. In view of Theorem 32.3, this definition agrees with the one given in the previous section. The reason for introducing this 'co-notation' is that we can now restrict ourselves to a single matroid M, without having to bring in M^*. To illustrate this, we prove the analogue for matroids of Theorem 9.3.

THEOREM 32.4. *Every cocycle of a matroid intersects every base.*

Proof. Let C^* be a cocycle of a matroid M, and suppose that there exists a base B of M with the property that $C^* \cap B$ is empty. Then C^* is contained in $E - B$, and so C^* is a cycle of M^* which is contained in a base of M^*. This contradiction establishes the result. //

COROLLARY 32.5. *Every cycle of a matroid intersects every cobase.*

Proof. Apply the result of Theorem 32.4 to the matroid M^*. //

From a matroid point of view, the two results in Theorem 9.3 are dual forms of a single result. Thus, instead of proving two results for graphs, as we had to in Section 9, it is sufficient to prove a single result for matroids and then use duality. Not only does this represent a saving of time and effort, it also gives us greater insight into several of the problems you met earlier in the book. One example of this is the similarity between the properties of cycles and cutsets. Another is a deeper understanding of duality in planar graphs.

As a further example of the simplifications introduced by matroids, let us return to Exercise 5.11. A straightforward proof of this result involves two separate proofs – one for cycles and a different one for cutsets. If, however, we prove the matroid analogue of the result for cycles, as stated in Exercise 30.7, then we simply apply it to the matroid $M^*(G)$ to deduce the corresponding result for cutsets. Conversely, we can use duality to deduce the result for cycles from the result for cutsets.

We now turn our attention to planar graphs, and show how the definitions of a geometric dual and an abstract dual of a graph arise as consequences of matroid duality. The Whitney dual of a graph, introduced in Exercise 15.11, is also a consequence of matroid duality, since the equation given in that exercise is simply a restatement of the expression for r^* at the beginning of this section.

We start with the abstract dual.

> **THEOREM 32.6.** *If G^* is an abstract dual of a graph G, then $M(G^*)$ is isomorphic to $(M(G))^*$.*

Proof. Since G^* is an abstract dual of G, there is a one–one correspondence between the edges of G and those of G^* such that cycles in G correspond to cutsets in G^*, and conversely. It follows immediately that the cycles of $M(G)$ correspond to the cocycles of $M(G^*)$. Thus, by Theorem 32.3, $M(G^*)$ is isomorphic to $M^*(G)$. //

> **COROLLARY 32.7.** *If G^* is a geometric dual of a connected plane graph G, then $M(G^*)$ is isomorphic to $(M(G))^*$.*

Proof. This follows immediately from Theorems 32.6 and 15.3. //

As remarked before, a planar graph G can have several different duals, whereas a matroid can have only one. The reason for this is that, if we have two (possibly non-isomorphic) duals of G, then their cycle matroids are isomorphic matroids.

We conclude this section by answering the question, 'under what conditions is a given matroid M graphic?' It is not difficult to find necessary conditions. For example, it follows from our discussion of representable matroids in Section 31 that such a matroid must be binary. Furthermore, by Exercise 31.10 and our discussion of the Fano matroid F, M cannot contain as a minor any of the matroids $M^*(K_5)$, $M^*(K_{3,3})$, F or F^*. It was shown by W.T. Tutte that these necessary conditions are sufficient. The proof of this result is too difficult to be given here (see Welsh [37]).

> **THEOREM 32.8 (Tutte 1958).** *A matroid M is graphic if and only if it is binary and contains no minor isomorphic to $M^*(K_5)$, $M^*(K_{3,3})$, F or F^*.*

On applying Theorem 32.8 to M^*, and noting that the dual of a binary matroid is binary, we obtain necessary and sufficient conditions for a matroid to be cographic.

> **COROLLARY 32.9.** *A matroid M is cographic if and only if it is binary and contains no minor isomorphic to $M(K_5)$, $M(K_{3,3})$, F or F^*.*

Tutte also proved that *a binary matroid is regular if and only if it contains no minor isomorphic to F or F^**. By combining this result with those of Theorem 32.8 and

Corollary 32.9, we immediately deduce the following matroid analogue of Kuratowski's theorem (Theorem 12.2).

THEOREM 32.10. *A matroid is planar if and only if it is regular and contains no minor isomorphic to $M(K_5)$, $M(K_{3,3})$ or their duals.*

Exercises 32

32.1^s   (i)   Show that the dual of a discrete matroid is a trivial matroid.

(ii)   What is the dual of the k-uniform matroid on a set of n elements?

32.2   Find the duals of the eight matroids on the set $E = \{a, b, c\}$, obtained in Exercise 31.2.

32.3   Verify property r(ii) of Section 30 for the function r^*.

32.4^s   Verify the result of Theorem 32.3 for the graph K_3.

32.5^s   What are the cocycles and cobases of

(i)   the 3-uniform matroid on a set of 9 elements?

(ii)   the cycle matroids of the graphs in Fig. 30.1?

(iii)   the cycle matroid of the graph in Fig. 31.2?

(iv)   the Fano matroid?

32.6   Give an example to show that the dual of a transversal matroid need not be a transversal matroid.

32.7   Show that the contraction matroid $M . A$ is the matroid whose cocycles are precisely those cocycles of M that are contained in A.

32.8*   Given that C is any cycle and C^* is any cocycle in a matroid, show that $|C \cap C^*| \neq 1$. (This is the generalization to matroids of Exercise 5.12.)

32.9*   Let M be a binary matroid on a set E.

(i)   Prove that, if M is an Eulerian matroid, then M^* is bipartite.

(ii)   Use induction on $|E|$ to prove the converse result.

(iii)   By considering the 5-uniform matroid on a set of 11 elements, show that the word 'binary' cannot be omitted.

(This result generalizes Exercise 15.9.)

33   Matroids and transversals

In the previous section, we illustrated the close connections between matroids and graphs. We now describe the connections between matroids and transversals. Our first aim is to simplify the proofs of some earlier results on transversal theory by taking a matroid point of view.

Recall that, if E is a non-empty finite set and $\mathcal{F} = (S_1, \ldots, S_m)$ is a family of non-empty subsets of E, then the partial transversals of $\mathcal{F}$ can be taken as the independent sets of a matroid $M(S_1, \ldots, S_m)$ on E. In this matroid, the rank of a subset A of E is the size of the largest partial transversal of $\mathcal{F}$ contained in A.

Our first example of the use of matroids in transversal theory is a proof of the result of Exercise 26.10, that a family $\mathcal{F}$ of subsets of E has a transversal containing a given subset A if and only if

(i) $\mathcal{F}$ has a transversal;

(ii) A is a partial transversal of $\mathcal{F}$.

It is clear that these conditions are necessary. To prove that they are sufficient, we observe that, since A is a partial transversal of $\mathcal{F}$, A is an independent set in the transversal matroid M determined by $\mathcal{F}$, and so can be extended to a base of M. Since $\mathcal{F}$ has a transversal, every base of M must be a transversal of $\mathcal{F}$, and the result follows immediately. If you have worked through Exercise 26.10, you will realize how much simpler this argument is.

Before showing how matroids can be used to simplify the proof of Theorem 27.3 on the existence of a common transversal of two families of subsets of a set E, we prove a matroid extension of Hall's theorem. Recall that, if $\mathcal{F}$ is a family of subsets of E, then Hall's theorem gives a necessary and sufficient condition for $\mathcal{F}$ to have a transversal. If we also have a matroid structure on E, then it is natural to ask whether there is a corresponding condition for the existence of an **independent transversal** – that is, a transversal of $\mathcal{F}$ that is also an independent set in the matroid. The following theorem, known as **Rado's theorem**, answers this question.

> **THEOREM 33.1** (Rado, 1942). *Let M be a matroid on a set E, and let $\mathcal{F} = (S_1, \ldots, S_m)$ be a family of non-empty subsets of E. Then $\mathcal{F}$ has an independent transversal if and only if the union of any k of the subsets S_i contains an independent set of size at least k, for $1 \le k \le m$.*

Remark. If M is the discrete matroid on E, then this theorem reduces to Hall's theorem, as stated in Theorem 26.1.

Proof. We imitate the proof of Theorem 26.1. As before, the necessity of the condition is clear. To prove the sufficiency, we show that if one of the subsets (S_1, say) contains more than one element, then we can remove an element from S_1 without altering the condition. By repeating this procedure, we eventually reduce the problem to the case in which each subset contains only one element, and the proof is then trivial.

It remains only to show the validity of this 'reduction procedure'. So, suppose that S_1 contains elements x and y, the removal of either of which invalidates the condition. Then there are subsets A and B of $\{2, 3, \ldots, m\}$ with the property that $r(P) \le |A|$ and $r(Q) \le |B|$, where

$$P = \bigcup_{j \in A} S_j \cup (S_1 - \{x\}) \quad \text{and} \quad Q = \bigcup_{j \in B} S_j \cup (S_1 - \{y\})$$

Then

$$r(P \cup Q) = r\left(\bigcup_{j \in A \cup B} S_j \cup S_1 \right) \quad \text{and} \quad r(P \cup Q) \ge r\left(\bigcup_{j \in A \cap B} S_j \right).$$

The required contradiction now follows, since

$$|A| + |B| \ge r(P) + r(Q)$$
$$\ge r(P \cup Q) + r(P \cap Q)$$
$$\ge \left| \bigcup_{j \in A \cup B} S_j \cup S_1 \right| + \left| \bigcup_{j \in A \cap B} S_j \right|$$
$$\ge (|A \cup B| + 1) + |A \cap B| \quad \text{(by Hall's condition)}$$
$$= |A| + |B| + 1. \; //$$

By imitating the proof of Corollary 26.2, we immediately obtain the following result.

COROLLARY 33.2 *With the above notation, F has an independent partial transversal of size t if and only if the union of any k of the subsets S_i contains an independent set of size at least $k + t - m$.*

We can now give a matroid proof of Theorem 27.3 on the existence of a common transversal of two families of subsets of a given set.

THEOREM 27.3. *Let E be a non-empty finite set, and let $\mathcal{F} = (S_1, \ldots, S_m)$ and $\mathcal{G} = (T_1, \ldots, T_m)$ be two families of non-empty subsets of E. Then $\mathcal{F}$ and $\mathcal{G}$ have a common transversal if and only if, for all subsets A and B of $\{1, 2, \ldots, m\}$,*

$$\left| \left(\bigcup_{i \in A} S_i \right) \cap \left(\bigcup_{j \in A} T_j \right) \right| \geq |A| + |B| - m.$$

Proof. Let M be the matroid whose independent sets are precisely the partial transversals of the family $\mathcal{F}$. Then $\mathcal{F}$ and $\mathcal{G}$ have a common transversal if and only if $\mathcal{G}$ has an independent transversal. By Theorem 33.1, this is so if and only if the union of any k of the sets T_i contains an independent set of size at least k, for $1 \leq k \leq m$ – that is, if and only if the union of any k of the sets T_i contains a partial transversal of $\mathcal{F}$ of size k. The result now follows from Corollary 26.3. //

We conclude with some results on the union of matroids. If $M_1, M_2, \ldots, M_k$ are matroids on the same set E, then we can define a new matroid $M_1 \cup M_2 \cup \cdots \cup M_k$, called their **union**, by taking as independent sets all possible unions of an independent set in M_1, an independent set in $M_2, \ldots$, and an independent set in M_k. The rank of this matroid is given by the following theorem, whose proof may be found in Welsh [37].

THEOREM 33.3. *If $M_1, \ldots, M_k$ are matroids on a set E with rank functions $r_1, \ldots, r_k$, then the rank function r of $M_1 \cup \ldots \cup M_k$ is given by*
   $r(X) = min \{r_1(A) + \cdots + r_k(A) + |X - A|\}$,
where the minimum is taken over all subsets A of X.

This result has the following useful corollaries.

COROLLARY 33.4. *Let M be a matroid. Then M contains k disjoint bases if and only if, for each subset A of E,*
   $kr(A) + |E - A| \geq kr(E)$.

Proof. M contains k disjoint bases if and only if the union of k copies of the matroid M has rank at least $kr(E)$. The result now follows immediately from Theorem 33.3. //

COROLLARY 33.5. *Let M be a matroid. Then E can be expressed as the union of k independent sets if and only if, for each subset A of E, kr(A) ≥ |A|.*

Proof. In this case, the union of k copies of the matroid M has rank $|E|$. It follows immediately from Theorem 33.3 that $kr(A) + |E - A| \geq |E|$, as required. //

If we apply these last corollaries to the cycle matroid $M(G)$ of a graph G, we easily obtain the following necessary and sufficient conditions for G to contain k edge-disjoint spanning forests, and for G to split into k forests. Since these results are not easy to obtain by more direct methods, we have thus again demonstrated the power of matroids in solving problems in graph theory.

THEOREM 33.6. *A graph G contains k edge-disjoint spanning forests if and only if, for each subgraph H of G,*

$$k(\xi(G) - \xi(H)) \leq m(G) - m(H),$$

where m(H) and m(G) denote the number of edges of H and G, respectively.

THEOREM 33.7. *A graph G splits into k forests if and only if, for each subgraph H of G, $k\xi(H) \geq m(H)$.*

Exercises 33

33.1[s]   Verify Rado's theorem when M is the Fano matroid, and $\mathcal{F} = (\{1\}, \{1, 2\}, \{2, 4, 5\})$.

33.2   Verify Corollary 33.4 when M is the 3-uniform matroid on a set of 8 elements.

33.3   Verify Corollary 33.5 when M is the 4-uniform matroid on a set of 9 elements.

33.4*   By modifying the Halmos–Vaughan proof of Hall's theorem, give an alternative proof of Theorem 33.1.

33.5[s]   Prove that a matroid M is a transversal matroid if and only if M can be expressed as the union of matroids of rank 1.

33.6   Dualize the results of Theorems 33.6 and 33.7 to obtain two further results in graph theory.

Appendix

This table lists the number of graphs and digraphs of various types with n vertices, for $n = 1, \ldots, 8$. Numbers greater than one million are given to one significant figure.

Types of graph $n =$	1	2	3	4	5	6	7	8
Simple graphs	1	2	4	11	34	156	1044	12 346
Connected simple graphs	1	1	2	6	21	112	853	11 117
Eulerian simple graphs	1	0	1	1	4	8	37	184
Hamiltonian simple graphs	1	0	1	3	8	48	383	6020
Trees	1	1	1	2	3	6	11	23
Labelled trees	1	1	3	16	125	1296	16 807	262 144
Simple digraphs	1	3	16	218	9608	$\sim 2 \times 10^6$	$\sim 9 \times 10^8$	$\sim 2 \times 10^{12}$
Connected simple digraphs	1	2	13	199	9364	$\sim 2 \times 10^6$	$\sim 9 \times 10^8$	$\sim 2 \times 10^{12}$
Strongly connected simple digraphs	1	1	5	83	5048	$\sim 1 \times 10^6$	$\sim 7 \times 10^8$	$\sim 2 \times 10^{12}$
Tournaments	1	1	2	4	12	56	456	6880

Bibliography

Of making many books there is no end;
and much study is a weariness of the flesh.
Ecclesiastes

Although we have almost reached the end of this book, we have by no means reached the end of the subject. We hope that you will wish to continue your study of graph theory, and for this reason, we suggest possible directions for further reading.

Other books at an introductory level are:

1. G. Chartrand, *Introductory Graph Theory*, Dover, 1985.
2. J. Clark and D. A. Holton, *A First Look at Graph Theory*, World Scientific Publishing, 1991.
3. F. Harary, R. Z. Norman and D. Cartwright, *Structural Models*, Wiley, 1965.
4. O. Ore, *Graphs and their Uses*, 2nd edn, New Mathematical Library 10, Mathematical Association of America, 1990.
5. R. J. Wilson and J. J. Watkins, *Graphs: An Introductory Approach*, Wiley, 1990.

Standard texts in graph theory include:

6. C. Berge, *Graphs*, North-Holland, 1985.
7. J. A. Bondy and U. S. R. Murty, *Graph Theory with Applications*, American Elsevier, 1979.
8. G. Chartrand and L. Lesniak, *Graphs & Digraphs*, 2nd edn, Wadsworth & Brooks/Cole, 1986.
9. F. Harary, *Graph Theory*, Addison-Wesley, 1969.
10. O. Ore, *Theory of Graphs*, American Mathematical Society Colloquium Publications XXXVIII, 1962.

A historical approach to the subject, including translations of many original sources, is given in:

11. N. L. Biggs, E. K. Lloyd and R. J. Wilson, *Graph Theory 1736–1936*, 2nd edn, Oxford, 1986.

Applications of graph theory, and the use of algorithms, are discussed in:

12. G. Chartrand and O. R. Oellermann, *Applied and Algorithmic Graph Theory*, McGraw-Hill, 1993.

13. N. Deo, *Graph Theory with Applications to Engineering and Computer Science*, Prentice-Hall, 1974.

14. A. K. Dolan and J. Aldous, *Networks and Algorithms: An Introductory Approach*, Wiley-Interscience, 1993.

15. S. Even, *Graph Algorithms*, Computer Science Press, 1979.

16. A. Gibbons, *Algorithmic Graph Theory*, Cambridge, 1985.

17. E. L. Lawler, *Combinatorial Optimization. Networks and Matroids*, Holt, Rinehart and Winston, 1976.

18. F. S. Roberts, *Discrete Mathematical Models, with Applications to Social, Biological and Environmental Problems*, Prentice-Hall, 1976.

19. M. N. Swamy and K. Thulasiraman, *Graphs, Networks and Algorithms*, Wiley, 1981.

20. A. Tucker, *Applied Combinatorics*, 2nd edn, Wiley, 1984.

21. R. J. Wilson and L. W. Beineke (eds.), *Applications of Graph Theory*, Academic Press, 1979.

Introductory books on combinatorics and transversal theory are:

22. I. Anderson, *A First Course in Combinatorial Mathematics*, 2nd edn, Oxford, 1989.

23. N. Biggs, *Discrete Mathematics*, 2nd edn, Oxford, 1993.

24. V. Bryant, *Aspects of Combinatorics*, Cambridge, 1993.

25. V. Bryant and H. Perfect, *Independence Theory in Combinatorics*, Chapman and Hall, 1980.

26. J. H. van Lint and R. M. Wilson, *A Course in Combinatorics*, Cambridge, 1992.

Specialist texts on some of the topics in this book are:

27. L. W. Beineke and R. J. Wilson (eds.), *Selected Topics in Graph Theory 1, 2, 3*, Academic Press, 1978, 1983, 1987.

28. S. Fiorini and R. J. Wilson, *Edge-Colourings of Graphs*, Research Notes in Mathematics 16, Pitman Publishing, 1977.

29. J. L. Gross and T. W. Tucker, *Topological Graph Theory*, Wiley-Interscience, 1987.

30. F. Harary and E. M. Palmer, *Graphical Enumeration*, Academic Press, 1973.

31. T. R. Jensen and B. Toft, *Graph Coloring Problems*, Wiley-Interscience, 1995.

32. J. W. Moon, *Counting Labelled Trees*, Canadian Mathematical Congress, 1970.

33. J. W. Moon, *Topics on Tournaments*, Holt, Rinehart and Winston, 1968.

34. J. G. Oxley, *Matroid Theory*, Oxford, 1992.

35. G. Ringel, *Map Color Theorem*, Springer-Verlag, 1974.

36. T. L. Saaty and P. C. Kainen, *The Four-Color Problem*, 2nd edn, Dover, 1986.

37. D. J. A. Welsh, *Matroid Theory*, Academic Press, 1976.

Sooner or later, you may need to refer to mathematical journals rather than to books. There are a large number of journals that are devoted to graph theory and related fields, such as the *Journal of Graph Theory*, the *Journal of Combinatorial Theory*, the *European Journal of Combinatorics*, *Ars Combinatoria*, and *Discrete Mathematics*.

Solutions to selected exercises

Chapter 1

1.1 (i) There are 5 vertices and 8 edges; vertices P and T have degree 3, vertices Q and S have degree 4, and vertex R has degree 2.

  (ii) There are 6 vertices and 5 edges; vertices A, B, E and F have degree 1 and vertices C and D have degree 3.

1.3 (i) Each carbon atom vertex has degree 4 and each hydrogen atom vertex has degree 1.

  (ii) The graphs are as follows:

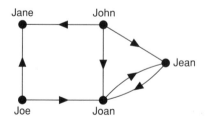

1.6 A suitable digraph is as follows:

Jane   John

Joe   Joan   Jean

Chapter 2

2.1 $V(G) = \{u, v, w, x, y, z\}$, $E(G) = \{ux, uy, uz, vx, vy, vz, wx, wy, wz\}$;
$V(G) = \{l, m, n, p, q, r\}$, $E(G) = \{lp, lq, lr, mp, mq, mr, np, nq, nr\}$.

2.3 (i) We can label the vertices as follows:

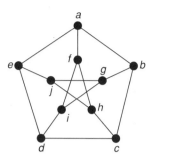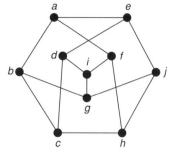

(ii) In the first graph, no vertices of degree 2 are adjacent; in the second graph they are adjacent in pairs. Since isomorphism preserves adjacency of vertices, the graphs are not isomorphic.

2.6 (i) graph 12; (ii) graph 27; (iii) graph 30.

2.7 graph 5: degree sequence (1, 1, 1, 3); sum of degrees = 6,   number of edges = 3;
graph 6: degree sequence (1, 1, 2, 2); sum of degrees = 6,   number of edges = 3;
graph 7: degree sequence (1, 2, 2, 3); sum of degrees = 8,   number of edges = 4;
graph 8: degree sequence (2, 2, 2, 2); sum of degrees = 8,   number of edges = 4;
graph 9: degree sequence (2, 2, 3, 3); sum of degrees = 10,   number of edges = 5;
graph 10: degree sequence (3, 3, 3, 3); sum of degrees = 12,   number of edges = 6.
In each case, the sum of the degrees is twice the number of edges.

2.10 The cycles with 5 and 6 vertices.

2.12
$$A = \begin{pmatrix} 0 & 1 & 0 & 0 & 1 \\ 1 & 0 & 1 & 0 & 1 \\ 0 & 1 & 0 & 2 & 0 \\ 0 & 0 & 2 & 0 & 1 \\ 1 & 1 & 0 & 1 & 0 \end{pmatrix} \quad M = \begin{pmatrix} 1 & 1 & 0 & 0 & 0 & 0 & 0 \\ 0 & 1 & 1 & 0 & 1 & 0 & 0 \\ 0 & 0 & 0 & 0 & 1 & 1 & 1 \\ 0 & 0 & 0 & 1 & 0 & 1 & 1 \\ 1 & 0 & 1 & 1 & 0 & 0 & 0 \end{pmatrix}$$

3.1

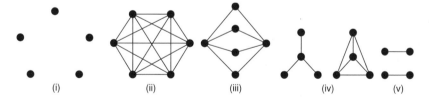

3.2 (i) 45; (ii) 35; (iii) 32; (iv) 14; (v) 15.

3.4 Regular graphs: 1, 2, 4, 8, 10, 18, 31;
bipartite graphs: 2, 3, 5, 6, 8, 11, 12, 13, 17, 23.

3.6 There are eight such graphs:

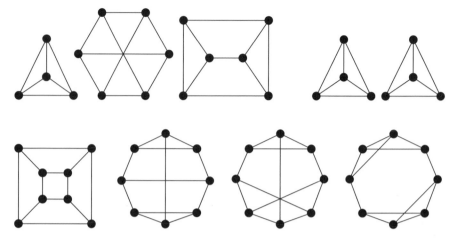

4.1   There are several possible solutions, all of them modifications of the solution in the text. One of these is as follows:

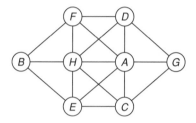

4.2   The following diagram illustrates such a gathering, with solid and dotted edges used as in the text.

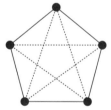

4.3   Using the method described in the text, we obtain the following graphs:

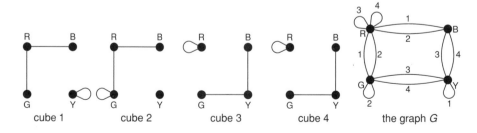

A pair of subgraphs H_1 and H_2 and a corresponding solution are as follows; there are several other solutions.

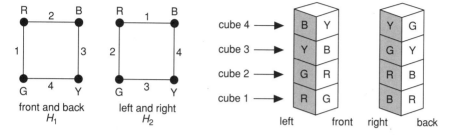

Chapter 3

5.1

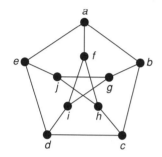

(i) $a \to b \to c \to d \to e \to j$;

(ii) $a \to b \to c \to d \to e \to j \to h \to f \to i \to g$;

(iii) $a \to b \to c \to d \to e \to a$; $a \to b \to c \to d \to i \to f \to a$;
$a \to b \to c \to d \to e \to j \to h \to f \to a$;
$a \to b \to c \to d \to e \to j \to g \to i \to f \to a$;

(iv) $\{ab, ae, af\}$, $\{ab, af, de, ej\}$, $\{ab, af, cd, di, ej\}$.

5.2 (i) 3; (ii) 4; (iii) 8; (iv) 3; (v) 4; (vi) 5; (vii) 5.

5.4 Let G be disconnected, and let v and w be vertices of G. If v and w lie in different components of G, then they are adjacent in $\bar{G}$. If v and w lie in the same component of G and z lies in another component, then $v \to z \to w$ is a path in $\bar{G}$. In either case, any two vertices can be connected by a path in $\bar{G}$, and hence $\bar{G}$ is connected.

5.5 (i) $\kappa = \lambda = 2$; (ii) $\kappa = \lambda = 3$; (iii) $\kappa = \lambda = 4$; (iv) $\kappa = \lambda = 4$.

6.1 (i) Eulerian; (ii) semi-Eulerian; (iii) neither; (iv) Eulerian; (v) neither.

6.2 Eulerian graphs: 1, 4, 8, 18, 21, 25, 31;
semi-Eulerian graphs: 2, 3, 6, 7, 9, 13, 14, 16, 17, 19, 22, 23, 26, 28, 30.

6.4 (i) At least $k/2$ trails are needed, so as to 'use up' all k vertices of odd degree. If we now add $k/2$ edges to G and join these vertices in pairs, then we obtain an Eulerian graph G'. We obtain the required $k/2$ trails by writing down an Eulerian trail for G' and then omitting the added edges.

(ii) Four.

6.5   There are many possible solutions; for example, traverse the edges in the order indicated by the following diagram:

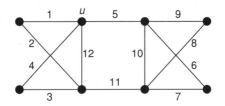

7.1   (i) Hamiltonian; (ii) semi-Hamiltonian; (iii) Hamiltonian; (iv) Hamiltonian; (v) Hamiltonian.

7.2   Hamiltonian: 1, 4, 8, 9, 10, 18, 22, 26, 27, 28, 29, 30, 31;
semi-Hamiltonian: 2, 3, 6, 7, 13, 15, 16, 17, 19, 20, 21, 23, 24, 25.

7.6 $K_{(n/2)-1,(n/2)+1}$, if n is even; $K_{(n-1)/2,(n+1)/2}$, if n is odd.

8.1   The permanent labels are, successively, $l(A) = 0$, $l(B) = 30$, $l(D) = 36$, $l(C) = 48$, $l(F) = 58$, $l(E) = 69$, $l(G) = 77$, and the shortest path, of length 77, is $A \to B \to D \to C \to F \to E \to G$.

8.5   Doubling the edges along the path $B \to D \to E \to A \to C$ yields a solution with total weight 24.

8.6   The required Hamiltonian cycle is $A \to B \to C \to E \to D \to A$, with total weight 14.

Chapter 4

9.1   The trees are 1, 2, 3, 5, 6, 11, 12, 13.

9.2

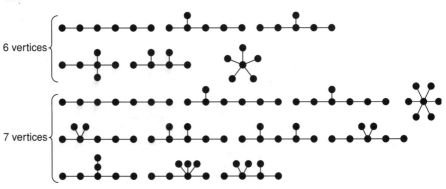

9.4

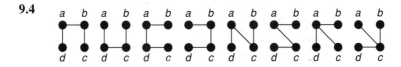

9.6   Cycles: $abcdea$, $abca$, $abcda$, cdc;
cutsets: $\{ab, ac, ad, ae\}$, $\{ac, ad, ae, bc\}$, $\{ad, ae, cd, cd\}$, $\{ae, de\}$.

9.8   (i) It is a bridge; (ii) it is a loop.

10.1   There are three unlabelled trees on 5 vertices.

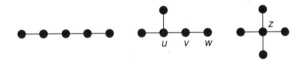

The first tree can be labelled in $(5!)/2 = 60$ ways; the second tree can be labelled
in $5 \times 4 \times 3 = 60$ ways, corresponding to the 60 possible choices for u, v and w;
the third tree can be labelled in 5 ways, corresponding to the 5 possible choices
for z. The total number is therefore $60 + 60 + 5 = 125$.

10.2   (i)

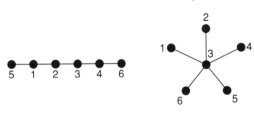

(ii) $(4, 4, 4, 1)$ and $(4, 2, 2, 4)$.

10.4

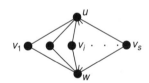

Each spanning tree in $K_{2,s}$ contains one of the two edges uv_i and v_iw, for each i,
together with one extra edge. The number of spanning trees is therefore
$2^s.s/2 = s2^{s-1}$.

11.1   We obtain either of the following weighted trees with total weight 13:

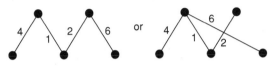

11.5   vertex A: $15 + (2 + 4) = 21$;   vertex B: $17 + (2 + 3) = 22$;
vertex D: $15 + (3 + 4) = 22$;   vertex E: $12 + (5 + 6) = 23$.

11.6   The graph is a connected graph with $n + (2n + 1) + 1 + 1 = 3n + 3$ vertices and
$\{4n + (2n + 1) + 2 + 1\}/2 = 3n + 2$ edges, and is therefore a tree, by Theorem
9.1(iii).

11.8   We obtain the following labelled trees, where the labels correspond to the order
in which the vertices are visited:

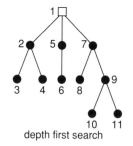

breadth first search   depth first search

11.10 The fundamental cycle equations are:

$VWZYXV$: $i_1 + i_3 - i_6 + i_7 = 12$; $VWZV$: $i_3 + i_5 + i_7 = 0$;
$VWZYV$: $-i_2 + i_3 - i_6 + i_7 = 0$; $WZYW$: $-i_4 - i_6 + i_7 = 0$.

The vertex equations are:

V: $i_1 + i_5 = i_2 + i_3$; W: $i_3 = i_4 + i_7$; X: $i_0 = i_1$; Y: $i_0 + i_6 = i_2 + i_4$; Z: $i_5 = i_6 + i_7$.

These equations have the solution

$i_0 = i_1 = 8, i_2 = 4, i_3 = i_4 = 2, i_5 = i_6 = -2, i_7 = 0$.

Chapter 5

12.1

(i)   (ii)

12.3 No, since $K_{3,3}$ is non-planar.

12.4 The complete graph K_n is planar if $n \leq 4$.
The complete bipartite graph $K_{r,s}$ $(r \leq s)$ is planar if $r = 1$ or 2, as shown by the following plane drawings:

 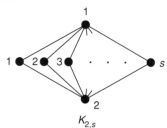

$K_{1,s}$ $K_{2,s}$

12.7 (i) and (ii)

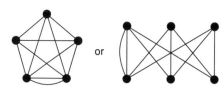

or

Although these graphs are not homeomorphic or contractible to K_5 or $K_{3,3}$, each contains a *subgraph* homeomorphic or contractible to K_5 or $K_{3,3}$.

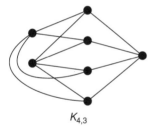   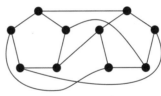

$K_{4,3}$ Petersen graph

12.10 The following drawings show the graphs drawn with two crossings. Some experimentation should convince you that no drawing is possible with just one crossing.

13.1   (i) $n = 8, m = 14, f = 8$, and $8 - 14 + 8 = 2$;
          (ii) $n = 6, m = 12, f = 8$, and $6 - 12 + 8 = 2$;
          (iii) $n = 9, m = 15, f = 8$, and $9 - 15 + 8 = 2$;
          (iv) $n = 9, m = 14, f = 7$, and $9 - 14 + 7 = 2$.

13.3   (i)   Since G has girth 5, we have $5f \le 2m$. Combining this with Euler's formula $n - m + f = 2$ gives the required inequality. If the Petersen graph were planar, then this inequality would be $15 \le 40/3$, which is false. Thus, the Petersen graph is non-planar.
          (ii) If G has girth r, then $rf \le 2m$. Combining this with Euler's formula gives the inequality $m \le r(n-2)/(r-2)$.

13.8   (i)   Since the Petersen graph is non-planar, its thickness is at least 2. But the Petersen graph can be obtained by superimposing two planar graphs, such as the outer pentagon and the 'spokes', and the inner pentagon. So the Petersen graph has thickness 2.
          (ii) Q_4 is not planar, as can be seen by applying Corollary 13.4(ii). So its thickness is at least 2. But Q_4 can be obtained by superimposing two planar graphs as follows.

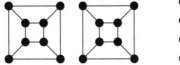   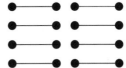

So Q_4 has thickness 2.

14.1

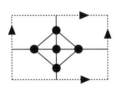   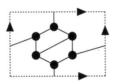

14.3   (i) $g(K_7) = \lceil (7-3)(7-4)/12 \rceil = 1$;
                $g(K_{11}) = \lceil (11-3)(11-4)/12 \rceil = \lceil 56/12 \rceil = 5$.
          (ii) K_8, as $g(K_8) = \lceil (8-3)(8-4)/12 \rceil = \lceil 20/12 \rceil = 2$.

14.5   (i)   the octahedron graph.

(ii)   For such a graph, $4n = 2m = 3f$. It follows from Theorem 14.2 that $m/2 - m + 2m/3 = 2 - 2g$, and so $m = 12(1 - g)$, which is non-positive. This contradiction shows that no such graph can exist.

15.1

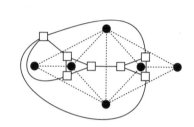

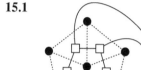

$n^* = f = 6, m^* = m = 10, f^* = n = 6, \quad n^* = f = 7, m^* = m = 11, f^* = n = 6.$

15.4   If such a plane graph existed, then its dual would be a plane graph with five mutually adjacent vertices. Since K_5 is non-planar, this is impossible.

15.5

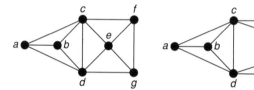

The above labelling shows that the given graphs are isomorphic. In the dual graphs the vertex-degrees are all 3 or 5 on the left and 3 or 4 on the right, so these duals cannot be isomorphic.

15.7   If G is a simple plane graph in which each vertex has degree 5 or 6, then G has at least 12 vertices of degree 5; if, in addition, each face is a triangle, then G has exactly 12 vertices of degree 5.

15.8   If G is 3-connected, then G has no vertices of degree 1 or 2, and hence G^* has no loops or multiple edges.

15.9   If G is bipartite, then each cycle of G has even length, and thus each cutset of G^* has an even number of edges; in particular, each vertex of G^* has even degree, and thus G^* is Eulerian. The reverse implication is obtained by reversing the argument.

16.1   (i)   An 'infinite star', obtained by joining the origin to infinitely many points on the unit circle;

(ii)   the complete graph with vertex set $\{x: 0 \leq x \leq 1\}$;

(iii)   the infinite hexagonal lattice;

(iv)   an infinite star, or an infinite path;

(v)   the graph obtained by adjoining an infinite path to a vertex of K_5;

(vi)   an infinite star, or an infinite path.

16.2   Consider the infinite star of Exercise 16.1(i).

Chapter 6

17.1   2 and 4.

17.3   2-chromatic: 2, 3, 5, 6, 8, 11, 12, 13, 17, 23;
3-chromatic: 4, 7, 9, 14, 15, 16, 18, 19, 20, 21, 22, 25, 26, 27, 29;
4-chromatic: 10, 24, 28, 30.

17.5   (i)   Upper bound = 3, chromatic number = 3;
(ii)   upper bound = k, chromatic number = 2.

17.7   If c_i is the number of vertices coloured i, for $1 \le i \le \chi\,(G)$, then $c_i \le n - d$. Thus, $n = c_1 + \cdots + c_n \le \chi(G)(n - d)$, and so $\chi(G) \ge n/(n - d)$.

19.2   Tetrahedron: 4; octahedron, 2; cube, 3; icosahedron, 3; dodecahedron, 4.

19.3   Any cycle graph with an even number of vertices; for example, C_4.

19.5   We prove the result by induction on the number of countries, the result being trivial for maps with at most six countries. Let G be a map with n countries, and assume that all maps with $n - 1$ countries are 6-colourable(f). By Euler's theorem, G contains a country F bounded by at most 5 edges. If we shrink F to a point, then the remaining graph has $n - 1$ countries, and is thus 6-colourable(f). A 6-colouring of the countries of G is then obtained by colouring F with a different colour from the (at most 5) faces surrounding F. Thus, G is 6-colourable(f).

20.1   4 and 3.

20.3   Chromatic index 2: 3, 6, 8, 13;
chromatic index 3: 4, 5, 7, 9, 10, 12, 15, 16, 17, 18, 20, 22, 23;
chromatic index 4: 11, 14, 19, 21, 24, 25, 26, 27, 28, 29.

20.4   (i)   Lower bound 2, upper bound 3, actual value 3;
(ii)   lower bound 7, upper bound 8, actual value 7;
(iii)   lower bound 6, upper bound 7, actual value 6.

20.6   Assume that $r \ge s$ and that $K_{r,s}$ is drawn as shown below, with the s vertices below the r vertices. Now successively colour the edges using the colours $\{1, 2, \ldots, r\}, \{2, 3, \ldots, r, 1\}, \ldots, \{s, \ldots, r, 1, \ldots, s - 1\}$.

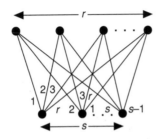

20.7   Since G is regular of degree 3, we have $\chi'(G) \ge 3$. To obtain a 3-colouring of the edges of G, we colour the edges of a Hamiltonian cycle alternately red and blue, and then colour the remaining edges green.

21.1   (i) $k(k-1)(k-2)(k-3)(k-4)(k-5)$;

(ii) $k(k-1)^5$.

K_6 can be coloured in $7.6.5.4.3.2 = 5040$ ways;

$K_{1,5}$ can be coloured in $7 \times 6^5 = 54\,432$ ways.

21.3   (i)

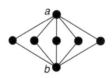

If vertices a and b have the same colour, then there are $k(k-1)^5$ colourings; if they have different colours, then there are $k(k-1)(k-2)^5$ colourings. Thus, $P_G(k) = k(k-1)^5 + k(k-1)(k-2)^5$.

(ii)   We have, by Theorem 21.1,

$$= k(k-1)^4 - k(k-1)(k^2 - 3k + 3) = k(k-1)(k^3 - 4k^2 + 6k - 4).$$

21.7   (i)   Since $k(k-1)^{n-1} = k^n - (n-1)k^{n-1} + \cdots + (-1)^{n-1}k$, G has n vertices, $n-1$ edges, and 1 component. It follows from Theorem 9.1(iii) that G is a tree on n vertices.

(ii)   Since $P_G(k) = k(k-1)^4$, G must be a tree on 5 vertices – that is,

Chapter 7

22.1   The first and last digraphs.

22.2   (i)   Since the underlying graph of D is connected, we have $n - 1 \leq m$, by Theorem 5.2. The upper bound is attained only by the digraph in which each pair of vertices is joined by two oppositely oriented arcs.

(ii) $n \leq m \leq n(n-1)$; the upper bound remains unchanged, and the lower bound follows since a cycle can be strongly connected, but no 'directed tree' can be.

22.3
$$\begin{pmatrix} 0 & 1 & 0 & 0 \\ 0 & 1 & 2 & 0 \\ 1 & 1 & 0 & 0 \\ 1 & 1 & 0 & 1 \end{pmatrix} \quad \begin{pmatrix} 1 & 0 & 0 & 0 & 0 & 0 \\ 0 & 0 & 0 & 0 & 0 & 1 \\ 0 & 1 & 0 & 1 & 0 & 0 \\ 0 & 0 & 0 & 0 & 1 & 0 \\ 0 & 0 & 1 & 0 & 0 & 0 \\ 1 & 0 & 0 & 0 & 1 & 0 \end{pmatrix}$$

22.6 G, 12; E, 10; B, 6.

23.1

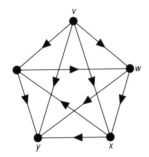

sum of out-degrees = 1 + 3 + 2 + 1 = number of arcs = 7;
sum of in-degrees = 1 + 3 + 3 + 0 = number of arcs = 7.

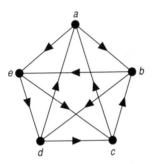

sum of out-degrees = 4 + 2 + 2 + 0 + 2 = number of arcs = 10;
sum of in-degrees = 0 + 2 + 2 + 4 + 2 = number of arcs = 10.

23.2   (i)   *aeda, edcbe, aecbda*;
(ii) *aedabdcbeca*;
(iii) *aecbda*.

23.3   If v and w are both sources, then vw and wv must both be arcs of the tournament, which is impossible; thus, no tournament can contain more than one source. The proof for sinks is similar.

24.1   (i)

$$\begin{pmatrix} 1 & 0 & 0 & 0 & 0 & 0 \\ \frac{1}{2} & \frac{1}{6} & \frac{1}{3} & 0 & 0 & 0 \\ 0 & \frac{1}{2} & \frac{1}{6} & \frac{1}{3} & 0 & 0 \\ 0 & 0 & \frac{1}{2} & \frac{1}{6} & \frac{1}{3} & 0 \\ 0 & 0 & 0 & \frac{1}{2} & \frac{1}{6} & \frac{1}{3} \\ 0 & 0 & 0 & 0 & 1 & 0 \end{pmatrix}$$

E_1 is persistent; all other states are transient.

(ii)
$$\begin{pmatrix} 0 & 1 & 0 & 0 & 0 & 0 \\ \frac{1}{2} & \frac{1}{6} & \frac{1}{3} & 0 & 0 & 0 \\ 0 & \frac{1}{2} & \frac{1}{6} & \frac{1}{3} & 0 & 0 \\ 0 & 0 & \frac{1}{2} & \frac{1}{6} & \frac{1}{3} & 0 \\ 0 & 0 & 0 & \frac{1}{2} & \frac{1}{6} & \frac{1}{3} \\ 0 & 0 & 0 & 0 & 1 & 0 \end{pmatrix}$$

All states are persistent.

24.2　(i)　Numbering the players clockwise around the table, we have:

$$\begin{pmatrix} \frac{1}{6} & \frac{1}{2} & 0 & 0 & \frac{1}{3} \\ \frac{1}{3} & \frac{1}{3} & \frac{1}{2} & 0 & 0 \\ 0 & \frac{1}{3} & \frac{1}{6} & \frac{1}{2} & 0 \\ 0 & 0 & \frac{1}{3} & \frac{1}{6} & \frac{1}{2} \\ \frac{1}{2} & 0 & 0 & \frac{1}{6} & \frac{1}{3} \end{pmatrix}$$

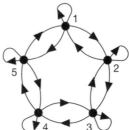

(ii)　By inspection, each state is persistent. Since $p_{ii} \neq 0$ for each i, each state is aperiodic. Hence the chain is ergodic.

Chapter 8

25.1　(i)

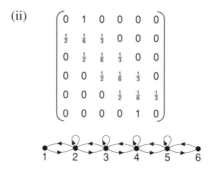

(ii)　*aw, bx, cy; aw, bz, cx; aw, bz, cy; ay, bz, cx; az, bx, cy.*

(iii)

A	$\varnothing$	a	b	c	ab	ac	bc	abc
$\lvert A \rvert$	0	1	1	1	2	2	2	3
$\lvert \varphi(A) \rvert$	0	3	2	2	4	4	3	4

Thus $\lvert A \rvert \leq \lvert \varphi(A) \rvert$ for each subset A of $\{a, b, c\}$.

25.3 The first, third and fourth vertices in V_1 are collectively joined to only two vertices of V_2, and hence the marriage condition fails.

26.1 (i) no transversal – the partial transversals are $\emptyset$, 1, 2, 3, 4, 5, 12, 13, 14, 15, 23, 24, 25, 35, 123, 124, 125, 134, 135, 234, 235, 1234 and 1235.

(ii) transversal – for example, $\{1, 2, 4, 5\}$.

(iii) no transversal – the partial transversals are $\emptyset$, 1, 2, 3, 12, 13, 23 and 123.

(iv) transversal – for example, $\{1, 4, 2, 5\}$.

26.3 By inspection, we can find eight transversals, each omitting just one of the eight letters in the word *MATROIDS*. For example, omitting M, we successively choose the letters S, R, O, I, D, A, T.

26.4 There is only one transversal – namely, $\{1, 2, \ldots, 50\}$.

26.6 (i) Let the sets in $\mathcal{F}$ be $S_1, \ldots, S_5$. Then the marriage condition fails for $\{S_3, S_4\}$ and $\{S_2, S_3, S_4\}$.

(ii) The union of any k of the subsets contains at least 1 element if $k = 1$ or 2, at least 2 elements if $k = 3$, at least 4 elements if $k = 4$, and 5 elements if $k = 5$, and so contains at least $k - 1$ elements for any value of k. But $t = 4$, $m = 5$, so $k + t - m = k - 1$, as required.

27.1

$$\begin{pmatrix} 1 & 2 & 3 & 4 & 5 & 6 & 7 & 8 \\ 2 & 5 & 8 & 3 & 6 & 1 & 4 & 7 \\ 3 & 6 & 1 & 5 & 2 & 7 & 8 & 4 \\ 4 & 7 & 6 & 8 & 1 & 3 & 5 & 2 \\ 5 & 8 & 2 & 7 & 3 & 4 & 1 & 6 \end{pmatrix}, \qquad \begin{pmatrix} 1 & 2 & 3 & 4 & 5 & 6 \\ 2 & 3 & 4 & 5 & 6 & 1 \\ 3 & 4 & 5 & 6 & 1 & 2 \\ 4 & 5 & 6 & 1 & 2 & 3 \\ 5 & 6 & 1 & 2 & 3 & 4 \\ 6 & 1 & 2 & 3 & 4 & 5 \end{pmatrix}$$

27.2

$$\begin{pmatrix} 1 & 2 & 3 & 4 & 5 \\ 5 & 3 & 1 & 2 & 4 \\ 2 & 1 & 4 & 5 & 3 \\ 3 & 4 & 5 & 1 & 2 \\ 4 & 5 & 2 & 3 & 1 \end{pmatrix}, \qquad \begin{pmatrix} 1 & 2 & 3 & 4 & 5 \\ 5 & 3 & 1 & 2 & 4 \\ 2 & 4 & 5 & 3 & 1 \\ 3 & 5 & 4 & 1 & 2 \\ 4 & 1 & 2 & 5 & 3 \end{pmatrix}$$

27.4 For both matrices, the term rank and μ are both equal to 4.

27.6 (i) $\{a, b, d\}$;

(ii) We check one case:
$|(\{c, e\} \cup \{a, b, d\}) \cap \{b, c, e\}| = |\{b, c, e\}| = 3$, and $3 \geq 2 + 1 - 3$.

28.1 edge form: the result is true with $k = 2$ for both graphs;
vertex form: the result is true with $k = 2$ for both graphs.

28.2 (i) and (ii) It is straightforward to verify that in each case there are exactly three disjoint paths joining any given pair of vertices.

28.4   The appropriate value of k is 3 for each graph.

29.1   (i)   The cuts are {*va*, *vb*}, {*va*, *ba*, *bd*}, {*va*, *ba*, *cd*, *dw*}, {*vb*, *ba*, *ac*}, {*vb*, *ba*, *cd*, *cw*}, {*ac*, *bd*}, {*ac*, *cd*, *dw*}, {*bd*, *cd*, *cw*} and {*cw*, *dw*}. The unique minimum cut is {*bd*, *cd*, *cw*} with capacity 6.

(ii)   A corresponding maximum flow with value 6 is as follows:

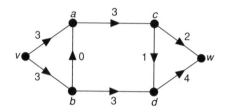

29.3   A cut with capacity 8 is {*BD*, *EG*, *EH*, *FH*, *FI*}.
A flow with value 8 is as follows:

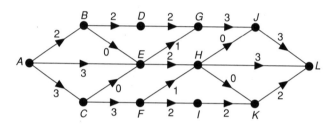

29.5   (i)   Add a new source vertex v^* joined to all the sources v_i with arcs v^*v_i of infinite capacity, and a new sink vertex w^* joined to all the sinks w_i with arcs w_iw^* of infinite capacity.

(ii)

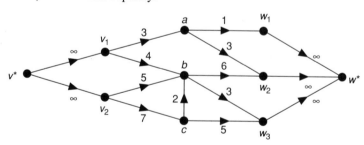

Chapter 9

30.1   (i)   Each subset of E is independent, there are no cycles, and $r(A) = |A|$ for each subset A of E; this is the *discrete matroid*.

(ii)   The only independent set is the empty set. The cycles are {*a*}, {*b*}, {*c*}, {*d*}, {*e*}, and the rank function is identically 0; this is the *trivial matroid*.

(iii)   The independent sets are those subsets of E with 0, 1, 2 or 3 elements, the cycles are those subsets with 4 elements, and if A is a subset of E, then $r(A) = \min \{|A|, 3\}$; this is the *3-uniform matroid*.

30.2 $M(G_1)$ has bases *abd, acd* and *bcd*; cycle *abc*; and independent sets Ø, *a, b, c,*
d, ab, ac, ad, bc, bd, cd, abd, acd and *bcd*.
$M(G_2)$ has bases *pqs, pqt, prt, pst, qrs, qrt, qst* and *prs*; cycles *pqr, rst* and *pqst*;
and independent sets Ø, *p, q, r, s, t, pq, pr, ps, pt, qr, qs, qt, rs, rt, st, pqs, pqt,*
prt, pst, qrs, qrt, qst and *prs*.

30.4 (i) The partial transversals are Ø, 1, 2, 3, 4, 5, 6, 12, 13, 14, 15, 16, 23, 24,
25, 26, 34, 35, 36, 123, 124, 125, 126, 134, 135, 136, 234, 235 and 236.
These partial transversals are the independent sets of the matroid $M(G)$,
where G is the following graph:

(ii) The bases are 123, 124, 125, 126, 134, 135, 136, 234, 235 and 236; the
cycles are 1234, 1235, 1236, 45, 46 and 56.

30.8 (i) This follows immediately from Exercise 5.11 and the definition of a
cutset.
(ii) graph G_1: *a, b* and *c*;
graph G_2: *pr, ps, pt, qr, qs, qt, rs* and *rt*.

31.1 Up to isomorphism, the four matroids on $\{a, b\}$ are:

bases	independent sets	cycles
Ø	Ø	*a, b*
a	Ø, *a*	*b*
a, b	Ø, *a, b*	*ab*
ab	Ø, *a, b, ab*	–

31.4 (i) Yes: with the notation of Exercise 30.2, $M(G_1) = M(ab, bc, d)$, $M(G_2) =$
$M(pq, qrs, st)$.
(ii) Yes: $M^*(G_1) = M(abc)$, $M^*(G_2) = M(pqr, rst)$.

31.6 If M is a k-uniform matroid on E, then $M = M(\mathcal{F})$, where $\mathcal{F}$ consists of k copies
of E.

31.8 The cycles of the Fano matroid are the lines, such as $\{1, 2, 4\}$, and the complements of the lines, such as $\{1, 2, 3, 6\}$.

32.1 (i) The only base of a discrete matroid on E is E itself, so the only base of its
dual is Ø; the dual matroid is thus the trivial matroid on E.
(ii) The $(n - k)$-uniform matroid on the same set of n elements.

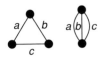

32.4   The bases for $M(K_3)$ are ab, ac, bc, so the bases of $(M(K_3))^*$ are a, b, c. Thus, $M(K_3^*)$ is isomorphic to $(M(K_3))^*$.

32.5   (i)   The cocycles are the subsets of cardinality 7; the cobases are the subsets of cardinality 6.

   (ii) $M(G_1)$ has cocycles ab, ac, bc, d, and cobases a, b, c.
   $M(G_2)$ has cocycles pq, prs, prt, qrs, qrt, st, and cobases pr, ps, pt, qr, qs, qt, rs, rt.

   (iii)   The cocycles are 1, 23, and the cobases are 2, 3.

   (iv)   The cocycles are 1236, 1257, 1467, 1345, 2347, 2456 and 3567, and the cobases are all subsets of four elements containing a line, such as 1247.

33.1   It is sufficient to note that $\{1, 2\}$ and $\{1, 2, 4, 5\}$ have ranks 2 and 3, respectively. An independent transversal of S is $\{1, 2, 5\}$.

33.5   If $M = M(S_1, \ldots, S_m)$ is a transversal matroid on E, then $M = M_1 \cup \ldots \cup M_m$, where M_k is the matroid on E whose bases are just the singleton subsets of S_k. Conversely, if $M_1, \ldots, M_m$ are matroids of rank 1, then their union $M = M_1 \cup \cdots \cup M_m$ is the transversal matroid on the sets $S_1, \ldots, S_m$, where S_k is the union of the independent sets in M_k.

Index of symbols

A	adjacency matrix	n	number of vertices
$A(D)$	arc-family of D	N_n	null graph
B	base of M	$P_G(k)$	chromatic polynomial of G
C_n	cycle graph	P_n	path graph
$\mathrm{cr}(G)$	crossing-number of G	Q_k	k-cube
D	digraph	r	rank function of M
$\deg(v)$	degree of v	r^*	rank function of M^*
E	non-empty finite set	$t(G)$	thickness of G
$E(G)$	edge set of G	T	tree
f	number of faces	u, v, w, z	vertices of G
F	Fano matroid	$v_0 \rightarrow \ldots \rightarrow v_m$	walk
g	genus	$V(D)$	vertex set of D
G	graph	$V(G)$	vertex set of G
$\overline{G}$	complement of G	W	cycle subspace of G
G^*	dual of G	$\widetilde{W}$	cutset subspace of G
$G(V_1, V_2)$	bipartite graph	W_n	wheel
$G_1 \cup G_2$	union of graphs	α, β, γ	colours
I	independent set of M	$\gamma(G)$	cycle rank of G
k	number of components	$\Gamma(G)$	automorphism group of G
K_n	complete graph	Δ	largest vertex degree of G
$K_{r,s}$	complete bipartite graph	$\kappa(G)$	connectivity of G
$K_{r,s,t}$	complete tripartite graph	$\lambda(G)$	edge-connectivity of G
$L(G)$	line graph of G	$\xi(G)$	cutset rank of G
m	number of edges	$\chi(G)$	chromatic number of G
M	matroid	$\chi'(G)$	chromatic index of G
M^*	dual matroid	$\mathcal{B}$	bases of M
$M \cdot A$	contraction matroid	C	cycles of M
$M \times A$	restriction matroid	$\mathcal{F}$	family of subsets
$M(G)$	cycle matroid	I	independent sets of M
$M(S_1, \ldots, S_m)$	transversal matroid		

Index of definitions

Only a little more
I have to write,
Then I'll give o' er
And bid the world Goodnight.
Robert Herrick

Here is my journey's end.
William Shakespeare (Macbeth)